REFLECTION, SCIENCE AND THE VIRTUES

CRITICAL ESSAYS ACROSS EDUCATION
Volume 2

Scope

This new book series aims at providing readers with a set of monographs dealing with current educational issues from a research and theoretical perspective. In dealing with the many problems besetting our increasingly globalised world, education remains one of the most critical professions, with educational research and theorising being one of the most potent vehicles for comprehending and informing future policies and practices. Thus *Critical Essays across Education* intends to provide academics, policy-makers, research students and concerned general readers, with critical reflections on up-to-date ideas from the international research field of education. In particular this series will reflect the growing trend for borderland crossings in education, whereby cross-discipline research is creating new and important theoretical pathways.

Much theoretical and research-based writing in educational texts tends towards the inaccessible end of the readability dimension. So the brief for intending authors in this series will be to reflect on their research, and those of others, in such a way as to help educate the generalist, as well as the specialist, readership

Reflection, Science and the Virtues

By

Tony Gibbons
University of South Australia

SENSE PUBLISHERS
ROTTERDAM / BOSTON / TAIPEI

A C.I.P. record for this book is available from the Library of Congress.

ISBN 978-90-8790-871-3 (paperback)
ISBN 978-90-8790-872-0 (hardback)
ISBN 978-90-8790-873-7 (e-book)

Published by: Sense Publishers,
P.O. Box 21858, 3001 AW Rotterdam, The Netherlands
http://www.sensepublishers.com

Printed on acid-free paper

TO EMILY AND TO JOHN KEEVES

TABLE OF CONTENTS

CHAPTER 1

STRUCTURE

Imre Lakatos wrote about the nature of science but his views have a further reach than that.

Lakatos provides the following account of his theory:

> Newtonian science, for instance, is not simply a set of four conjectures – the three laws of mechanics and the law of gravitation. These four laws constitute only the "hard core" of the Newtonian Programme. But this hard core is tenaciously protected from refutation by a vast "protective belt" of auxiliary hypotheses. And, even more importantly, the research programme also has a "heuristic", that is, a powerful problem-solving machinery, which, with the help of sophisticated mathematical techniques, digests anomalies and even turns them into positive evidence. (Lakatos, 1978b, p.4)

The "heuristic" is both positive and negative. The negative heuristic is that the hard core may not be rejected or modified. The positive heuristic solves problems and provides for the progress and development of the research programme. While it continues to do this, the research programme is progressive. When it fails to do this the research programme is degenerating. A research programme degenerates when auxiliary hypotheses are added ad hoc in the attempt to defend the hard core. In principle, no scientific theory, which constitutes the hard core, can be disproved as it is always possible to save it with an auxiliary hypothesis. The other side of the coin, that no scientific theory can ever be proved because it is always underdetermined by available evidence (Hempel, 1966), is undermined by Lakatos. Lakatos stipulates that the reasonableness of induction must be presupposed. I shall argue that this presupposition can be justified.

What Lakatos calls a 'research programme', Kuhn calls a 'paradigm'. For Kuhn (1967) science proceeds by scientists accepting and working within a paradigm to solve the puzzles that present themselves. Puzzles that appear insoluble within the structure of the paradigm are deemed 'anomalies'. There is a critical stage at which the range and importance of the anomalies create a disturbance within the paradigm and for its practitioners such that a revolution occurs. The revolution is the production and acceptance of a new paradigm. The new paradigm is acceptable because, at least, it covers that which the old paradigm explains and also explains the anomalies.

There are a number of points that can be drawn out of the ideas of Lakatos and Kuhn.

– There is a hard core of propositions to any research programme or paradigm which determines what are to count as reasons and acceptable argument. A change to a new hard core or paradigm results in a change in what are to count as reasons and acceptable argument. It is a change to the way in which experience is perceived and accounted for. It is thus a change in rationality. The notion of a 'hard core' is a commonplace in human thinking and belief in many areas, as are the determined attempts to defend it.

- While there is a continuous attempt to keep the research programme or paradigm coherent there will be anomalies. Incoherence is the mark of all research programmes or paradigms for scientific ideas have to be assessed as the best possible so far. Again, the matter extends beyond science. In all fields the human condition is one of fallibility. Human inquiry is always subject to incoherence, inconsistency and fault. It is a mistake to abandon the search for the truth because of this human condition.
- Scientists tend not to abandon a research programme or paradigm unless there is a better one to replace it. At what point this occurs is moot. At some point the incoherence of the research programme or paradigm leads to a search for a new research programme or paradigm. Of itself, this will not produce change. Normally, there must be a new research programme or paradigm which deals with all that the previous one did and solves at least a major portion of the anomalies.

Thus, the crucial element in falsification is whether the *new theory* offers any novel, excess information compared with its predecessor and whether some of this excess information is corroborated (Lakatos, 1978b, p.36) .

The point must be phrased as a tendency, for human psychology, social circumstances and other factors than the quest for truth are involved. Again, the point extends beyond science. The hard core of a person's belief about reality is not changed easily. That hard core is part of their identity as a person. The implication of the above is that it is possible to compare paradigms which, in turn, raises the problem of commensurability. The implication of what has been said so far is that there is no independent rationality to which the problem may be referred. There is, however, reality.

At some point we would be justified to say of an individual scientist that he or she was behaving irrationally if they continued to stand by a degenerate and discarded research programme or paradigm. This might well have been said of those who refused to accept the plate theory of the world after paleomagnetic evidence was available in the late 1950s. However, the point at which a charge of irrationality might be made must depend upon the incoherence of the old paradigm, the rationality of the new paradigm and the extent of its acceptance as providing a new way of seeing the old and dealing with anomalies. Again, this point applies to the whole of human experience. History provides numerous examples of people adhering tenaciously to beliefs which others claim are irrational. Human beings can change, however.

Lakatos and Kuhn were writing about the nature of scientific research. MacIntyre writing of science says:

> The criterion of a successful theory is that it enables us to understand its predecessors in a newly intelligible way. It, at one and the same time, enables us to understand precisely why its predecessors have to be rejected or

> modified and also why, without and before its illumination, past theory could have remained credible. It introduces new standards for evaluating the past. It recasts the narrative which constitutes the continuous reconstruction of the scientific tradition. (MacIntyre, 1977, p.460)

Both Lakatos and Kuhn approached their work through the history of science. MacIntyre develops this approach by describing science as a continuous narrative. Science has to be seen as a continuous attempt to understand and explain the world of our experience. At any particular stage in the narrative of science there will be theory, belief, myth and stories. Part of this will be the hard core. A grasp of the narrative is essential to the understanding of the present. Einstein cannot be understood except in the context of Newton. Copernicus cannot be understood without Ptolemy. Science is not a series of isolated events in which understanding jumps from one hard core to the next. It is a continuous narrative which is not a linear progression.

Because each change in the narrative produces a change in the rationality of science, that is, what is to count as acceptable argument, relevant evidence, good reasons, there appears to be a relativism. Certainly, the science of a particular time or culture is relative to a paradigm and to the culture that accepts that paradigm. It is relative to the current hard core. However, changes can occur as the narrative proceeds and these changes are towards increased understanding and knowledge. I say towards for it is clear that there have been blind alleys and mistakes. But the notion of a blind alley or a mistake presupposes that there is the possibility of success and progress. That scientific theory is to be deemed the best so far is not to say that it is relative in the sense of having no connection with the truth. To suppose relativism is to give up the search for knowledge; it is to forget that all science is about the attempt to understand and explain the world of our experience and, that, ultimately, all science is tested against that world. This is to distinguish between the rationality of the hard core and the test for truth. The rationality of the hard core is internal to it. The test for truth is the correspondence of that hard core to reality. It is therefore possible to recognise a hard core as being relative to a time and a culture while at the same time see it as a step along the road to understanding reality.

This is not simply an account of science, it is an account of rationality as such. There is no reason to suppose that it is an account that is peculiar to science. MacIntyre (1977) sees no reason to confine the ideas to science. He is of the view that it is applicable to ethics and attempts to provide a research programme in that area (MacIntyre, 1981, 1988, 1990a, 1999).

I am of the view that it is applicable to any research programme, paradigm or narrative that attempts to explain and provide understanding of the world of our existence. It is true of the study of history, geography, literature. It is true of any culture for a culture, by definition, is an attempt to deal with the world of our existence. For instance, in the eighteenth century the arrival of Europeans in Australia placed in conflict two very different cultures. The ontology and epistemology of the two cultures were radically different. The ways in which they saw and analysed the world were very different. There were two types of rationality which can be compared. One culture can be judged superior if it can be shown that the incoherence in one culture is overcome by the other and that there are advances in ontology and epistemology. However, in making such a judgement it must be remembered that the change is only to the best theory so far.

It may be that in response to contact with another culture there is a modification to the hard core of a culture. It is often a mistake to think that modifications to the hard core can be confined to a particular area. Consider Western Europe pre- and post- Copernicus. The belief that the Earth was the centre of the planetary system found expression not only in astronomy but also in art, literature and social activities. With the change to a belief in a solar-centred system there was an effect on astronomy but art, literature and social activities also changed.

A revolution in thought of perhaps greater impact occurred when Western Christian Europe gained access to the writings of Aristotle in the twelfth century. There was a profound effect on the existing hard core throughout Western Europe.

This brief account of Lakatos and Kuhn serves a number of purposes. It makes a point about the human condition, that we tend to have a hard core which encompasses our view of reality and we are loath to change it. We raise defences and these defences can become ad hoc. It raises the problem of understanding the ontological views of others. What is it to understand such a thing? Is it possible or are views untranslatable? It raises the problem of criticising other points of view when that view is so basic to their identity. How does one criticise and decide for or against what might seem to be incommensurable? Can there be independent criteria to which we can refer? All of these problems raise their heads in a consideration of the nature of science and, importantly, what it is to learn science. But, importantly, these issues are not confined to science. Whether or not it was his intention Lakatos picked out something which applies to the human condition. Science is not an insulated study and, if in what follows the arguments are about science it should not be supposed that they are not about human thought and the human quest for knowledge in what Oakeshott called the conversation of mankind.

CHAPTER 2

UNDERSTANDING

The work of sociologists in the sociology of knowledge owes a great deal to the work of Weber. For Weber (1949) social phenomena involved the actions of persons who attach meaning to what they are doing. In order to study the actions of such agents the sociologist must develop the technique or method of *Verstehen*. Without this method the sociologist is not in a position to do anything. Weber is not the originator of this method. That honour belongs to Dilthey, whose approach to the study of human beings in society has been summarised as:

1. All human manifestations are part of a historical process and should be explained in historical terms. The state, the family, even man himself cannot be adequately defined abstractly because they have different characteristics in different ages.

2. Different ages and differing individuals can only be understood by entering imaginatively into their specific point of view, what the age or the individual thought relevant must be taken into account by the historian.

3. The historian himself is bound by the horizons of his own age. How the past presents itself to him in the perspectives of his own concerns becomes a legitimate aspect of the meaning of that past. (Rickman, 1972, p.405)

Das Verstehen is a method that attempts to identify and describe the meaning that people give to their lives or to particular instances in their lives. It is a method that can be applied to the individual, to groups, to the culture, to the age. In each case it is a search for meaning. On the basis of this meaning it is then possible to give a complete explanation of what has occurred. This presupposes that the meaning can be rendered in terms which can be understood. Translation is an issue.

The work of scientists is a social phenomenon. Scientific theories exist in a social context and are the product of the work of scientists. For Dilthey, in order to explain the work of scientists the researcher must turn to the method of *Verstehen* and seek the meaning that scientists give to their work. The current sociologists of knowledge are following in this tradition. What counts as a creditable and justifiable scientific theory is determined by social factors and to elucidate these social factors the sociologists must enter into the point of view of the age in which the theories are produced and applied and enter into the point of view of the scientists working with those theories. Only then is it possible to explain why a theory came into existence and why it is considered creditable and justifiable.

There are problems with this approach if the technical meaning of understanding is confused with normal usage.

The history of philosophy contains a number of investigations of the concept of understanding usually in the context of a consideration of the difference between knowledge and understanding. The positions occupied are many and varied ranging from Anselm's *Credo ut intelligam* through Kant's categories to Hegel's distinction between dialectical thought, reason, and pre-dialectical thought.

In the face of such disparate views it is worth considering how we use the word. It is certainly the case that we use the verb 'to understand' to indicate that meaning has been perceived, that an idea has been grasped, that something has been comprehended. However, we also use the word in other ways which are linked to this meaning but distinct as in Wittgenstein's family of meanings.

I can understand English but not the language of my forbears, Irish. In understanding a language I am able to express information about myself and the world and to understand the expression of similar information by others. Therefore, one sense of 'understand' is to be able to grasp the meaning of the words, that is, to grasp what it is to which the words refer. This is the public reference of language. It is therefore possible to recognise a word and its area of reference but not to understand it in a complete sense. For instance, I may be able to recognise that some 40 different words in the language of the Inuit refer to 40 different states of snow but I do not have the understanding which enables me to grasp the particular reference for I cannot see the snow as they do. This is linked to the notion of fluency in language. However, even though a person may be fluent in two languages he or she still may not be able to translate successfully all the terms in one language into another. Terms can carry with them cultural baggage which militate against this. How is 'simpatico' translated into English? These remarks are clearly relevant to the problem of translation which has arisen previously.

If, as a barrister, I say that I understand the criminal law then I am asserting that I am thoroughly familiar with the criminal law and that I apprehend clearly the character and nature of the criminal law. Notice, however, that if, as a lay person, I say that I understand the criminal law I would not be taken to be claiming the understanding that a criminal barrister has. In this case it could well be that I would be asserting that I can see the point or significance of the criminal law and particular parts of it. The distinction between the two usages appears to lie in the breadth and depth of knowledge which is ascribed.

It is common to use 'understand' to indicate some knowledge. So we talk of understanding that X whereby we indicate that X is a settled fact. We talk of coming to understand whereby we indicate that we have acquired knowledge or learnt something or heard something. In all these usages we understand about a matter. We say that we understand why one thing follows from another. For instance, I may say, on learning that water expands as it approaches freezing point, that I now understand why pipes burst in cold weather. The understanding here is an understanding of the relationship between one empirical incident and another. It is this sort of understanding that one often finds in science.

We also talk of the understanding as an apparent reference to the exercise of intelligence and judgement. A person who understands possesses and shows intelligence and judgement. We talk of understanding in the sense of sympathy and tolerance.

All the above may seem to indicate that understanding is a synonym for knowledge. Perhaps on occasion, propositional knowledge, perhaps on occasion knowing how, perhaps on occasion knowing why. What then do we make of the sentence; 'I know why he or she did it but I cannot understand why he or she did it?' It would seem that when we say that we are indicating that we are unable to put ourselves in the other's shoes. We cannot see the situation as he or she sees it and, perhaps, we doubt that he or she sees the situation in its entirety. In any event Dilthey was correct to insist that *Verstehen* was a technical approach which was not to be confused with the ordinary use of the term understanding. Certainly, 'understanding' in one of its senses is used as indicating that one has grasped the meaning of something, and understanding can also refer to sympathetic awareness of another's viewpoint. However, this is only part, and not the main part of the use of understanding.

When the sociologists of knowledge insist, as do Barnes and Bloor, that what counts as a creditable and justifiable scientific theory is determined by social factors and to elucidate these social factors the sociologists must enter into the point of view of the age in which the theories are produced and applied and enter into the point of view of the scientists working with those theories, then they are confusing the methodology which is useful in examining the history of scientific pursuits with the credit of the theory that is produced. To understand why a person sees it so is not to understand why it is so. To understand why a person sees it is so is not to know that it is so. It is only to see it how they see it. It is another question as to whether or not it is correct to see it that way.

CHAPTER 3

SOCIOLOGY OF KNOWLEDGE

The views in the sociology of knowledge range from the radical to the progressive. The radical being someone who asserts that knowledge is entirely the construct of society while the progressive is one who argues that knowledge is partially constructed by society but that the natural world has some effect. There are of course an infinity of positions between the radical and the progressive depending upon the amount that the natural world enters into the equation.

The claim is that the physical sciences are subject, in part or totally, to definition, control, dissemination and manipulation by the societies in which they exist. There is no denying that there is a partial influence, the radical claim is that of total or determining influence and is the more interesting. This claim questions the traditional categories which have grown up in epistemology. In particular, it calls into question the traditional notion of 'truth' and the traditional notion of 'knowledge' as justified true belief. In consequence, it threatens the notion that there is a world of reality independent of the person which can be known or attempted to be known by intellectual pursuits of which the sciences are but one group.

To respond to this claim by arguing that it is a version of the genetic fallacy serves no useful purpose. The claim applies equally to philosophy and epistemology which are themselves controlled, disseminated and organised by these same groups. Philosophy, in particular, is looked upon as an elite intellectual pursuit though more so in some societies than others. The elite in the social group that is philosophers control entry to the group and manipulate its research programs.

Scientists live and work in a social milieu. They themselves are the products of particular cultures and particular upbringings. It would be absurd to doubt that the work of scientists is influenced by such social factors. The question is whether or not the social factors affect the epistemological status of their work and, if so, to what extent. Studies in the history of science in recent years have led to an emphasis on the way in which social factors affect the work of scientists. As might be expected it has also led some to claim that scientific knowledge is not objectively justifiable. One version of the claim is that what is called scientific knowledge is the result of a variety of social factors which are both external and internal to the work of the scientist. Such a claim does not deny that there might be objectively true scientific theories or false theories. Nor is it a claim that what is counted as objective scientific knowledge depends on the standards of the group. It is the claim that what constitutes a credible or justifiable theory is determined by social factors which have no connection with the empirical data that the scientist may be considering. A theory may be considered justifiable and accepted as such.

It may also be either true or false in a sense which is objectively independent. The scientists do not determine its acceptance because of independent evidence rationally considered but because of social influences to which they are subject. This is the strong program in the sociology of knowledge initiated by Barnes and Bloor. Barnes and Bloor (1982) accept that, within the practice of science, rational scientific practice may be used to limit the number of possible theories. But, after that process, the acceptance of one theory over another is determined by social factors. The effect this would have on science teaching is evident. Barnes and Bloor emphasize that scientific theories are underdetermined by the evidence and that, therefore, it is necessary to go to social factors in order to find the cause of acceptance of theory. Sokal and Bricmont give a succinct explanation of underdetermination:

> Another idea, often called the "Duhem-Quine thesis", is that theories are underdetermined by evidence. The set of all our experimental data is finite; but our theories contain, at least potentially, an infinite number of empirical predictions … there is always a large (even infinite) number of theories compatible with the data – and this, whatever the data and whatever their number. (Sokal & Bricmont, 1998, p. 69)

Sokal and Bricmont point out that the Duhem-Quine thesis is often used to pick out the notion that all observations are theory-laden and also that, in their opinion, Duhem's view of the issue is less radical than Quine's. However, in common parlance, the Duhem-Quine thesis is often used to refer to the underdetermination of theories by the evidence.

The problem with the use of the Duhem-Quine thesis by Barnes and Bloor is that there is no indication that they see any distinction worth making between competing theories on the grounds of the weight of evidence. Yet they admit rational culling may take place. A theory may have little or no supporting evidence. On the other hand it may have a great deal of supporting evidence. A theory may have little supporting evidence but, what there is, is relevant and remarkable. A theory may have a great deal of supporting evidence but that evidence is of little weight. If we accept that all theories are underdetermined by the evidence and that science is a matter of pursuing the truth without, perhaps, ever quite catching it, then the practice of science becomes a matter of weighing the evidence and making a decision on that basis. To do otherwise would be to treat all evidence, of whatever type or description or relevance, as of equal weight. This flies in the face of scientific practice and, since Barnes and Bloor must rely on historical accounts of scientific practice to support their case, they find themselves in a position in which their argument undermines itself.

Barnes and Bloor summed up their approach by a description of the merits of the strong program. Their statement of the strong program for the investigation of knowledge set the scene for much that followed. According to Barnes and Bloor the investigation of knowledge should abide by a number of propositions. Couvalis sets out these propositions in a way which illuminates the approach.

> 1. It seeks *causal* explanations of *all* scientists' beliefs – explanations in which psychological and other components play a part but in which a social component plays the crucial part.
>
> 2. It is *impartial* with respect to the truth and falsity, rationality and irrationality, of beliefs.
>
> 3. It gives *symmetrical* explanations of beliefs in the sense that the same *types* of causal explanations are given for all beliefs.
>
> 4. It is *reflexive* because the sort of explanation which advocates of the strong program give of the credibility of scientific theories can, in principle, be applied to explaining the credibility of the strong program. (Couvalis, 1997, p.144)

Their approach has been defended in Bloor (1981) and Barnes and Bloor (1982).[1]

The immediate problem lies in the meaning to be attributed to the four propositions. Clearly the propositions are intended to apply to both the content and the justification of knowledge. Some indication of this is given in Barnes and Bloor (1982). Here they assert that 'true' and 'false', 'rational' and 'irrational' function as local referents. That is, their meaning is locally determined. In this manner the problem of contradiction between beliefs held in different societies is eliminated. What counts as rational is also locally determined.

> For the relativist there is no sense attached to the idea that some standards or beliefs are really rational as distinct from merely locally accepted as such. (Barnes & Bloor, 1982, p.27)

The traditional approach to rational and irrational, which is to argue that the first is explained by giving reasons and the second is explained by some species of sociological or psychological explanation, is rejected. Both rational and irrational are to be explained by reference to social factors. These assertions immediately create a problem for the methodology that they propose. Unless they can justify the strong program remaining outside of, or common to, all cultures then it becomes local and cultures outside the one for which Barnes and Bloor are writing have no cause to accept it. It cannot be outside all cultures for this would be to place it in the position of being an arbiter to which cultures would have to turn. It would become a species of independent criteria against which the local would be measured. However, since all theory is socially caused, the notion of criteria independent of culture seems inconsistent. There appear to be two solutions. One is to claim that, as a matter of fact, all cultures hold this method in common. This would require the support of empirical research; in the case of science, research in the history of science. Even a brief encounter with the history of the work of Aristotle, Galileo, Newton, Darwin and Boltzmann is enough to undermine the proposition that all cultures hold the methodology of the strong program in common. A second solution is to claim that all cultures should hold the method in

[1] Kukla (2000) discusses the attacks by Laudan (1981) and adds his own criticism.

common. But on what grounds? If it is because it is more reasonable to do so than not, then, this is incompatible with the whole project of the strong program. If it is because of something to do with some social values then again we are back with empirical research and the attempt to show that societies hold some values in common. If the grounds are social factors then the whole process is faced with infinite regress and the problems that brings with it.

The strong program betrays some incoherence but it serves to draw attention to the social factors, which include social values, that influence science and to undermine the individualistic view which has held sway for so long. Couvalis (1997) makes a number of criticisms of the strong program which appear to me sound. However, I take issue with him on his account of the relationship between social factors and the practice of science. He says, setting out the arguments of those he calls rationalists:

> … producing social explanations of how scientists come to have beliefs on scientific matters is not incompatible with rationalism. Some rationalists explain how scientists come to have such beliefs by arguing that the scientific community is structured so that social factors and the personal interests of scientists are likely to induce them to arrive at their beliefs on the best evidence available. (Couvalis, 1997, p. 148)

It is not that this view is incorrect but that it is only a partial, and the less important part at that, of an account of social factors that affect scientific belief. Couvalis accepts these ideas and in summing up the merits of the strong program says:

> The claim that science is importantly a social enterprise is plausible. What tends to make scientists work well is the training and the constraint and reward structure of the scientific community. It typically limits the degree to which prejudices and social agendas can be used to interpret data and to draw inferences. In addition, good scientific work is always done in a social structure by scientists who usually defer to authorities and rely on the work of others in a systematic way. (Couvalis, 1997, p. 151)

I shall argue that it is the social and human structure that contains within it factors that lead to objectively independent scientific ideas and that Couvalis has missed the constitutive element present in the social and human structure of science. I shall also argue that these social elements are not idiosyncratic to science but are operative in the same way in the community within which science exists. First, however, it is important to consider some developments since the strong program, specifically the work of Longino.

CHAPTER 4

LONGINO'S CONSTRUCTIVISM

Initially there is the strong program with its emphasis on the content and justification of knowledge being the product of social factors. Alongside this are the works of Latour (1987; 1992) and Knorr-Cetina (1983), which emphasise the social interaction between scientists as they practise science and the effect that this has on scientific theory and belief. As might be expected others have moved on from the argument that social factors determine what is credible and justifiable scientific work by focusing on particular social factors. Gender is one such social factor that has attracted some attention. Values are another. The position adopted in these areas is in contrast with that taken by Barnes and Bloor (1982). For Barnes and Bloor it may be possible to ameliorate the effect of social factors but it is impossible to eradicate them. Scientific work is always determined by social factors, which include values, of one sort or another. Those working in the field of gender have tended to take the view that scientific work is objectively justifiable if the effect of this particular social factor can be eliminated. This is an understandable approach given the issue involved. There is an unwillingness to believe that gender discrimination cannot be eliminated. Some of those working in the field of values have taken a different approach. For them the values of a scientist may be justifiably operational in scientific work. So it would be justifiable for a scientist to have regard to some results on the basis of his or her values and to discard others (Longino, 1990). Longino has also written extensively on gender issues.

Longino has said of her work:

> My project has been to develop an analysis of scientific inquiry that both acknowledges the social dimensions of inquiry and keeps room for the normative and prescriptive concerns that have been the traditional preoccupation of philosophers. (Longino, 1994a, p.135)

The works of Barnes and Bloor she takes to be concentrating on the practice of science within the scientific community and arguing that it is the social interests of practitioners that determine whether or not a theory or belief is accepted in the scientific community. The correspondence between the model and the external world is not the determining factor. As I have noted Barnes and Bloor must rely, and do rely, on empirical studies in the history of science to support their case. Longino sees the weakness of this approach in that case studies in the history of science cannot show that scientists had no other reason but their social interests in making their decisions.

Latour and Knorr-Cetina she takes to be putting forward a more compelling thesis. Taking Knorr-Cetina's (1983) fundamental question to be – How is that which we come to call knowledge constituted and accepted? – she argues that the question predetermines the answer in favour of Knorr-Cetina's approach. It is a question that is aimed at the actual practice of science which will give no account of the way in which science ought to be practised. It is however an interesting question because it can lead us to see how the practice of science may have been undermined or distorted or biased, not least of all in favour of the practice of male scientists. Clearly, if Knorr-Cetina argues that this or that account of a scientific episode reveals distortion or bias then there is a presupposition that there are norms against which the episode may be judged. It is the norms that interest Longino.

Longino has produced a sophisticated account of the effect of social factors on scientific practice in which she moves on from the work of Barnes and Bloor, Latour and Knorr-Cetina. Longino argues from the underdetermination of scientific theories by the evidence available to the possibility of a number of theories fitting the evidence. This is an unremarkable move. It is, of course, a move that raises the question of how the scientist is to decide between the possibilities. One answer is that it cannot be done in any rational manner. Another answer is that it is socially determined. Longino's answer is that scientists always have recourse to some background assumptions, which underpin their decisions. These background assumptions may be internal or external to the science. Internal background assumptions are necessary to connect the data to the theory. One such background assumption is a principle of induction. However, Longino points out that this principle is problematic (Longino, 1990). Longino's solution is to turn to values which enable a decision to be made on the theory to adopt.

Longino raises the question of values and science and, necessarily, the question of whether or not science can be value-free. At this point it is necessary to make some distinctions for there are a number of uses of 'value-free' in relation to science.

- 'Value-free' in relation to science is often used to indicate that inferences that proceed from 'is' to 'ought' commit a fallacy. In consequence the only values to which a scientist should subscribe are those internal to the practice of science. I shall consider this so-called fallacy later.
- 'Value-free' can be an assertion that there are values internal to scientific practice which are not reducible to other non-scientific values. Both Longino and Allchin (1999) appear to hold a species of this view.
- 'Value-free' can be used to mean that scientific practices are not subject in any way to values outside science, to social, political and religious values for instance. Longino (Longino, 1994b) explicitly rejects this view.

A paper by Allchin provides a common view of the relation of science and values:

> The common characterization of science as value-free or value-neutral can be misleading. Scientists strongly disvalue fraud, error and 'pseudo-science', for example. At the same time, scientists typically value reliability, testability, accuracy, precision, generality, simplicity of concepts and heuristic power. Scientists also value novelty …. The pursuit of science as an activity is itself an implicit endorsement of the value of developing knowledge of the material world. (Allchin, 1999, p. 2)

Allchin considers the above to be the ultimate values of science and notes that Merton identified several institutional norms or imperatives that apply to the practice of science. Among these Allchin refers to organized scepticism, disinterestedness and communism. These norms Allchin refers to as proximate or mediating values and claims that they follow from the ultimate values.

Allchin continues:

> While the pursuit of scientific knowledge implies a certain set of characteristically 'scientific' values, the relevance of other values in the practice of science are not thereby eclipsed. Honesty is as important in science as elsewhere, and researchers are expected to report authentic results and not withhold relevant information … Science is not exempt from ethics or other social values. (Allchin, 1999, p. 3)

He argues that the ends of research and the practice of the research are both subject to social values and he poses the situation in which scientific values are followed but social values are infringed as in the case of some Nazi research. His distinction between scientific values and social values is a common view.

Longino has much in common with Allchin. Longino asserts that:

> According to a standard conception of the relation between science and values, ethical issues are in an important sense external to the sciences. While those points of contact between science and the values of the social and cultural context in which it is done may determine the directions of research or of its applications, within the boundaries so determined scientific inquiry itself proceeds according to its own rules. (Longino, 1994b, p. 139)

This is the same distinction that Allchin makes and, as Longino notes, is the orthodox position.

Longino argues that the values to which scientists turn may be internal or external. Internal values are those values which are constitutive of scientific practice. External values are contextual, that is, they exist in the society or culture in which the scientific practice takes place.

Constitutive values may include injunctions to seek the explanation which is the simplest, the most fertile, the most elegant. There is argument over what may constitute the list of such internal values but there is no doubt that they exist. Contextual values may include racial, spiritual, economic or any of the wide range of values present in a particular society. There is no doubt that, on occasion, contextual values are used to determine both the direction of scientific research and also which theory is acceptable. Longino's thesis is that it is legitimate, on occasion, to use contextual values to determine which theory is acceptable. She supports her case by arguing that Newtonian mechanics was accepted in preference to rival theories because it fitted a mechanistic view of the world of the culture in which Newton lived (Longino, 1990). Couvalis states her argument as

> … the mathematical treatment of nature, which was part of the mechanistic picture, enabled people to control certain aspects of nature better than did

> rival accounts. Overall, then, the mechanistic picture of nature suited the values and needs of craftspeople and the rising capitalist class, whereas its rivals did not (Longino, 1990, pp. 92-98; Couvalis, 1997, p. 154)

Barnes and Bloor would not disagree. On Longino's account, it then became the case that the rationality of the culture changed with the values and needs of an expanding powerful group. This affected science and affected what was seen to count as good science. In other words, the constitutive values of science changed under the effect of the contextual values. Clearly this can occur. However, if the constitutive values of science become a result of contextual values, then the objectivity of science becomes problematic. It could be argued that this is precisely what happened to science in Nazi Germany and Stalin's Russia.

Longino argues that the problem of the loss of scientific objectivity is avoided if scientists function as a community with shared values. Longino (1990) had argued that the practice of science is a social activity and it is this social activity that underpins the objectivity of scientific practice. Following this she turned to an examination of the production of scientific knowledge.

> Here [as distinct from Longino (1990)] I would like to concentrate on a different sort of argument, namely, conceptual arguments to the effect that observation and justificatory reasoning – central elements of scientific knowledge construction – are themselves social. (Longino, 1994a, p. 139)

Longino (1994a) refers to a paper by Knorr-Cetina and Amann (1990) which provided evidence that social interactions between scientists led to increasingly definite accounts of the data observed. This, she claims, is supported by observations of other instances of scientific practice. This, of course, depends upon the empirical evidence, but is there anything in the concept of observation which can lead to the conclusion that social interaction is a necessary feature of observation? Her argument is twofold. First, there must be agreement as to what is to count as what. Is this an acid or not, does this result fall within the parameters or not? Second, it is a requirement of the practice of science that experiments be repeatable and that therefore:

> The requirement of the repeatability of experiments is a requirement of the intersubjective accessibility of data serving as evidence because it is a requirement that anyone similarly placed with similar equipment would see (perceive) the same thing. Nature has been enrolled as an ally when the experimenter has found a way to stabilize her or his results across different perspectives. (Longino, 1994a, p. 140)

With regard to reasoning, Longino asserts:

> The second main cognitive element in the production of scientific knowledge is reasoning, which traces lines of evidentiary support between data and theories and hypotheses. Individual brains or minds, of course engage in calculation. But reasoning (at least scientific reasoning) is not mere calculation; it is bringing the appropriate considerations to bear on

> judgments. Justificatory reasoning is part of a practice of challenge and response: challenge to a claim is met by the offering of reasons to believe it, which reasons can then be challenged on grounds both of truth and of relevance, provoking additional reasoning. (Longino, 1994a, p. 141)

Longino is arguing that there are social norms which operate within the practice of science which, if followed, lead to increasing consensus on scientific ideas. The social norms are directed to ensuring that scientific data and ideas are given wide circulation so that they may be discussed and criticised. This procedure leads to justifiable scientific results which can be claimed to have as much as objectivity as is possible. These social norms are constitutive values of science and are learned as a person is initiated into the practice of science and are the result of scientific consensus and social interaction. They are background assumptions on which a scientist is entitled to rely and are to a large extent invisible in the practice of science though they may be publicly scrutinised at times. I take issue with the way in which Longino confines herself to reasoning for there is something more than reasoning involved here, reflection.

Longino's view of the scientific community and its work is given in the following propositions.

> (1) There must be publicly recognized forums for the criticism of evidence, of methods, and of assumptions and reasoning.
>
> (2) There must be uptake of criticism. The community must not merely tolerate dissent; its beliefs and theories must change over time in response to the critical discourse taking place within it.
>
> (3) There must be publicly recognized standards by reference to which theories, hypotheses, and observational practices are evaluated and by appeal to which criticism is made relevant to the goals of the inquiring community. With the possible exception of empirical adequacy, there needn't be and probably isn't a set of standards common to all communities. The general family of standards from which those locally adopted are drawn includes such cognitive virtues as accuracy, coherence, and breadth of scope, and such social virtues as fulfilling technical or material needs or facilitating certain kinds of interactions between a society and its material environment or among the society's members. The point of requiring public standards is that, by explicitly or implicitly professing adherence to those standards, individuals and communities adopt criteria of adequacy by which they may be evaluated. The satisfaction of goals of inquiry is not ascertained privately, but by evaluation with respect to shared values and standards. This evaluation may be performed by anyone, not just be members of the community sharing all standards. Furthermore, standards are not a static set, but may themselves be criticized and transformed in reference to other standards, goals, or values held temporarily constant. Indeed, as in the case of observation and the assumptions underlying justificatory reasoning the presupposition of reliance on such standards is that they have survived critical scrutiny.

> (4) Finally, communities must be characterized by equality of intellectual authority. What consensus exists must be the result not of the exercise of political or economic power, or of the exclusion of dissenting perspectives, but a result of critical dialogue in which all relevant perspectives are represented. This criterion is meant to impose duties of inclusion; it does not require that each individual, no matter what her or his past record or state of training, should be granted equal authority on every matter. (Longino, 1994a, pp. 144-145)

The above are, for Longino, the features of an ideal community which is practising the pursuit of knowledge, knowledge communities. It is through the work of this ideal community that science retains its claim to objectivity. She conceives of a general family of standards which include cognitive and social virtues and it is from this general family that different knowledge communities draw their idiosyncratic set.

Throughout the paper quoted the discussion is concerned with scientific knowledge. It would seem fair to take all mention of community to be referring to a scientific community. In which case Longino is espousing the notion that there are different scientific communities with different sets of standards, the set being not static. The set is drawn from a general collection of standards which include virtues, both cognitive and social, and presumably at different times and for different purposes different virtues may be called upon to constitute the set.

If, on the other hand, we read Longino as using community to refer to any community whose practice is the pursuit of knowledge then this could include mathematicians, artists, theologians, and philosophers. While this may seem a fairer way of reading Longino, the situation is still the same with respect to the set of standards by which any community conducts its practice. Philosophers may be considered as communities of linguistic analysts, Aristotelians, Kantians or Hegelians all pursuing knowledge in their own way within their own communities and having little more than acerbic contact with other communities. Each community would then have its own set of virtues by which it conducted its practice.

But it is doubtful if it is possible to take this interpretation of Longino. Longino follows Grene (1966, 1985) in taking science to be a social practice but departs from Grene in disagreeing with the application of MacIntyre's (1981) concept of a practice to science. Longino discusses the practices of science and portrait painting, as MacIntyre does, and comes to different conclusions. For Longino, science, as distinct from portrait painting, is social and complex. This complexity is to be found in the many different types of people who are part of the social group that come together in a particular scientific inquiry. She names technicians, research associates, the experimenter and mathematical modeller and others. These people form the group that work together towards a particular end and participate in the inquiry at different stages and different levels. Knowledge, she asserts is not the particular possession of any one of these people. Evidently she thinks portrait painting to be a singular endeavour which is essentially non-social in character and essentially non-complex in character. Presumably she would apply the same critique to writing, musical composition and other practices where it might seem to be the case that a single person is involved. It is hardly necessary to point out the mistake of identifying and characterising a practice by reference to what appears to be the main practitioner. The mistake in science is with the concentration on singular figures like Galileo and Newton. It is the same mistake to think of the practice of portrait painting simply in terms of a Rembrandt or a Goya or the practice of musical composition in terms of a Bach or a Mozart. The social part, the cooperative human activity of portrait painting does not depend upon persons interacting socially and cooperatively at a given instance but on the meaning given to their activity through the development of painting over a long period of time. In this way portrait painting is a socially established and socially constituted form of human activity. Longino, in emphasising the current social circumstances of an activity, neglects the meaning given to a practice through its narrative.

A number of issues have arisen in discussion of Longino's work.

- The problematic nature of induction is central to Longino's argument. I shall argue that induction is reasonable.
- A distinction has been made between scientific values and values external to science. I shall argue that any distinction that picks out scientific values and, as a separate group, social values, is mistaken and a distortion of the practice of science and a distortion of the role of the scientist in the community in which he or she exists.
- Longino has listed a number of virtues as being part of the ideal community of scientists. I shall argue that there is a necessary link between the practice of science and the virtues.

CHAPTER 5

THE SCIENCE CURRICULUM

The emphasis on the way in which social factors affect what is to count as scientific knowledge leads inevitably to the notion that there may be a multitude of sciences, each of which reflect the culture in which they arise and in which they are viable. The National Science Teachers Association (NSTA) of the USA in 2000 adopted as one of its tenents:

> Curricular content must incorporate the contributions of many cultures to our knowledge of science. (Association, 2000)

The problem is to determine what the incorporation of cultural contributions may mean. For instance, in line with the NSTA there are statements such as:

> The nature and practice of science builds on traditions of observation and inquiry found in many cultures. Viewing experiences, ideas and phenomena through the lenses of diverse cultural sciences provides a breadth and depth of understanding that is not possible from any one cultural perspective. (South Australian Department of Education, 2003)

As a general comment on the ways in which human beings have approached the world of their experience in an attempt to understand it this is not contentious. It is indeed salutary for a human being in one culture to realise that people in other cultures attempt explanations of the physical world in radically different ways. This assumes that the views of one culture can be effectively expressed in terms people from another culture can understand. The study of other cultures is a matter of ethnography and a significant area of study but it is not physics or any of the other natural sciences. However, there is an implication that explanations of the physical world generated by one culture are limited and that a greater breadth and depth of understanding and explanation are available to those who study the way other cultures deal with the same phenomena. This is a confusion. In Western culture there is an explanation of the visible colours based on a theory of light and refraction. If, in another culture, there is an explanation of the same phenomenon based on the spirits then this is incompatible with the first explanation and, in combination, does not give rise to a greater depth and breadth of understanding of the phenomenon. It does give rise to a depth and breadth of understanding of human thought and culture but that is not the physics of optics.

The nature and practice of science in a culture certainly is built on the narrative history of the practice in that culture. It is an expression of the rationality of that culture. However the matter is frequently taken further by assertions such as:

> Every culture has its own ways of thinking and its own world views to inform its science. Western science is the most dominant form of science but it is only one form among the sciences of the world. Through conjoint thinking and the identification of boundaries between ways of knowing, values and insights provided by the sciences of diverse cultures can be mutually supportive. (South Australian Department of Education, 2003)

The use of the term 'dominant' suggests that the pre-eminent position of Western science is a socio/political matter rather than an epistemological one. Dominance is a term that indicates a power relationship. The assertion is that every culture "has its own ways of thinking" and the implication is that there are cultural sciences. However, what you call science I may call witchcraft. In both cases one of two things is presupposed. Either an independent, justifiable account of what is a science is presupposed, or, a particular culture presupposes its account of the practice of science as correct and judges accordingly. Both accounts seem to presuppose that it is possible for a person from one culture to understand another culture. Then, building on this presupposition, it is presupposed that it is possible to compare and judge, in the first case against independent criteria, in the second case against the criteria of a culture. There is a third possibility – cultures using the name 'science' to identify different practices and the practices are incommensurable. In fact, the first two accounts presuppose independent criteria. For where the science of a culture is judged according to the criteria of a second and deemed not to be a science, then, it is presupposed by that second culture that it has a set of criteria which have a truth value and can be seen as justifiable by an independent observer. The matter therefore resolves itself into a question of whether or not the sciences of different cultures are translatable, and, if so, whether or not this renders them commensurable or incommensurable. On one reading of the above quotation it would appear that the position taken is that the sciences of different cultures are translatable and incommensurable.

We have here the appearance in the sciences of the debate which has raged elsewhere. Are there independent criteria against which the merits of a work of art, literature, drama can be judged or is it matter of what is good in its time and its place? Are there independent criteria against which moral action can be judged or is it matter of subjective assessment by the individual in his or her particular social place. the quotation may also be read as asserting a privileged, spectatorial position. In the case of the sciences it may be held that the spectator is armed with criteria by which a science may be recognised and, using these criteria, may survey the various cultures of the world and be able to judge whether or not a culture has a science. On this reading it does not follow that if a culture calls a practice a 'science' then that is what it is. This implies that there is some description of the practice of science against which the practices of various cultures are examined. The presupposition is that it is possible to translate the science of another culture sufficiently well to be able to understand the claims that are being made. Some may be suspicious that the spectator is looking through Western eyes or, more specifically, the eyes of the Western ethnographer. As Barba remarks about the conventional Western approach to science teaching:

> For all children, learning science involves an assimilation to a Eurocentric/androcentric way of knowing and learning. (Barba, 1993)

And there are suggestions in South East Asian societies that Western science carries with Western baggage. If this is so then the judgement of that spectator may be biased on two counts. But, putting this to one side, what might constitute the criteria or criterion which a spectator may use? It might be supposed that the purpose served by a practice marks it as a science. The spectator then looks to see whether any practice in a culture serves that purpose. If it does so, then that culture possesses a science. Horton (1971) argues that the importance of theory and causal explanations may be thought of as a distinguishing mark of Western science. The production of theory and causal explanations of the world of our experience is the purpose of the practice of science. However, he argues, traditional societies attempt precisely the same. African traditional societies construct theory by recourse to spirit gods and place things in a causal context by connecting these spirit gods with the social context in which the person concerned exists. Thus both Western cultures and traditional cultures possess sciences though clearly they may be completely incompatible in many respects. This may be thought to be the case in countries where Western science and Aboriginal science exist within the same country.

Science curricula typically define science in terms of the exercise of rationality and assert that there is a defining characteristic of science that is independent of a culture – it is rational inquiry. It is necessary to remark that rational inquiry applies to many other intellectual pursuits, it is not exclusive to science. There are problems if the notion of a rational inquiry is set alongside the notion that all cultures have their own ways of thinking. Rationality in science in a particular culture may be conceived of as being constituted by shared assumptions and standards of evidence. That is, the hard core of beliefs that lie at the centre of science in a particular culture determine what is to count as evidence, reasons and sound argument and explanation in science in that culture (Lakatos, 1978a). Different cultures may well have ontological and epistemological presuppositions which define their culture and also define their science.

There are indications here of links to Barnes, Bloor, and Longino who propose to define the practice of science in social terms. For Barnes and Bloor (1982), what constitutes a credible scientific theory is determined by social factors which have no connection with the scientist's empirical data. For Longino (1990), social factors outside the practice of science are background assumptions that determine the outcome of scientific practice. In both cases there is the notion that science is a social construction. In both cases the traditional notion of 'truth' is under attack.

A particular culture may have a dictum – when seeking to explain a phenomenon always refer to the spirit world, or, when seeking to explain a phenomenon always seek a causal explanation. This amounts to an injunction to be observed in solving problems and presupposes particular views of the world of our experience and particular views of what is to count as knowledge, evidence and sound argument and explanation. This is part of the rationality of that culture. There is a hard core or cores, which determine the scope of the practice and what is to count as evidence, reason and sound argument and explanation. This hard core is composed of both scientific ideas and cultural ideas. Surrounding the hard core is a vast array of auxiliary hypotheses erected to defend and preserve the core (Chalmers, 1976; Lakatos, 1978a; Couvalis, 1997).

At one time the belief in Western Europe was that the sun revolved around the earth. The system was geocentric and is usually referred to as the Ptolemaic system. Ptolemy was not the originator of the system but was responsible for drawing together all that was known in the Mediterranean world of the second century A.D.. This geocentric system was accompanied by ontological and epistemological beliefs that were empirical, historical, religious and mythic. Eventually the matter took on social and political significance as well. This hard core of belief with accompanying hypotheses determined what was to count as evidence, reason, sound argument and explanation. It determined the rationality of the tradition in the context of that culture. The tradition viewed the sun and the planets from the point of view of an observer on the earth. Copernicus asked himself what would the view be like from somewhere else, the sun for instance. He published his theory of a heliocentric system in 1543 and died the same year. As Bronowski remarks:

> To us the place of Aristotle's mechanics and Ptolemy's astronomy in the medieval system seems incidental. But to the contemporaries of Copernicus, they represented the natural and visible order of the world. The wheel as the Greek ideal of perfect motion had become a petrified god, as rigid as the Mayan calendar or the figures carved on Easter Island.
>
> The system of Copernicus seemed unnatural to his age, even though the planets still run in circles. (Bronowski, 1973, p.197-198)

The hard core of a tradition defines the 'natural and visible order of the world'. It defines the rationality. The change proposed by Copernicus was a change in rationality and seemed unnatural. In this example it is possible to see that rationality is particular to a tradition and to a culture; that it may be difficult and, at times, impossible for another culture to comprehend that rationality; that the rationality can change, sometimes as in a revolution, at other times in a slow process. It is possible to acknowledge of the diversity of scientific views and the cultures in which they reside but to conclude from the diversity and the consequent diversity of rationalities that no further steps are possible is to assume that translation and comparison are not possible. It is to move from the view that each science is a social construction of that society's view of reality to the view that there can only be recognition and acceptance. There is an unwillingness to make ontological and epistemological judgements about other cultural sciences and an unwillingness to propose a notion of truth that lies outside the particular culture. However, the touchstone is always reality and the search for truth about that reality.

While the rationality may be relative, truth is not. The purpose of science is to seek explanations of the physical world. Proposed explanations are tested against the physical world and, depending upon their success in accounting for that physical world, may be accepted as a step in the search for truth. The matter is a search, a quest, for the condition of scientific inquiry is one in which there is progress from one theory to the next. The current theory is always the best theory so far. This is the fallible human condition and applies not just to science but to all human inquiry (Hempel, 1966; MacIntyre, 1977). As a result it is possible to conceive of comparing the merits of disparate cultural sciences. The first step is to examine the internal consistency of a science and the second step is to examine its account of the physical world. It may then be possible to say that one cultural science is better than another.

Clearly the notion of incommensurability must be examined further. This, necessarily will involve the discussion of other concepts peculiarly important in science – theory, values, models and experiments. It will also involve arguments about the possibility of translation.

INCOMMENSURABILITY

It is clearly possible to argue as follows. There are a myriad of non-Western views of the world. And, in any event, there are within Western European society a large range of world views some of which are incompatible with each other. However, I shall continue to use the phrase Western science for the moment. Traditional societies may have a rich and detailed knowledge of the plants and animals that inhabit their part of the earth and which are important to their survival. Cultures may have a rich and detailed knowledge of astronomy and have viable means of navigation. To this may well be added a variety of myths, beliefs and rituals which serve the society. And in all this it is tempting to see a science which performs the same function in its culture as Western science and the culture is none the worse for it. Science, on this account, is a practice which performs a certain function in a culture and, provided it performs that function successfully then there is no cause to claim that it is worse than the science of another culture. If the science sustains the culture then how is it possible to say that, for instance, Western science is a superior form of science? The assertion is that Western science is only one form of science. At present it is the dominant form of science. Other cultures have different forms of science. Sciences have to be evaluated by reference to the cultures which they serve. The science particular to a culture should be taught and the sciences of other cultures may be taught for cross-cultural understanding. The practice of science in any culture is a socially established cooperative human activity. It is coherent according to the criteria imposed on it within that culture. It depends for its continued survival on sufficient shared beliefs and procedures and an ability to cope with novel issues. The shared beliefs may be about spirits, they may have their source in myths or they may be the beliefs that we associate with Western science. The problem then becomes the problem of comparing cultural practices.

In which case, given that the dominant Western form of science is taught in schools, are other forms of science introduced:

– for the purposes of comparison?

- for the purposes of cross-cultural understanding?
- for the purposes of showing why Western science is the dominant form?
- for the purposes of selecting and using other sciences in preference to the dominant form? or,
- for the purposes of providing optional areas of study?

A multiplicity of views exist and the matter is best viewed as a spectrum. Matthews (1994, ch. 9) considers the problem of multicultural science education and provides an account of the problems involved.

It is a mistake to think that there is no method by which the problem of incommensurability can be tackled. Consider Western science and the example of an existing scientific paradigm facing attack from within, as classical mechanics was from Boltzmann. Classical mechanics was an account of the world of our experience which had a measure of internal coherence and of correspondence with reality which had satisfied scientists for decades. The first step in assessing a paradigm is always to attempt to show that by the internal standards of the tradition the paradigm is incoherent. This Boltzmann did. The second step is to attempt to show that a new proposal solves the incoherence and gives an account of why it was bound to occur. This Bohr did. Note that the second step involves giving a narrative of the paradigm for it came into existence originally to solve certain problems which had presented themselves. A history of a paradigm is, on this view, an essential part of understanding the paradigm, its weaknesses and whether or not a solution has presented itself. Note also that the step involves showing why the problem was bound to occur. This goes to an assessment of the weakness of the paradigm and provides a justification for seeking a solution. A third step is not essential but would consist in starting to consider what incoherencies are bound to occur within the new proposal. The presupposition here is that while the final end of the inquiry may be the truth, the pursuit of the inquiry will yield only steps towards that final end and these steps will necessarily contain flaws which hopefully will be corrected (Einstein and Infeld, 1967). This serves to direct the future of the inquiry. Why is it thought that this methodology cannot be followed when considering the practice of science within different cultures? It is essentially a rational methodology and should be applicable. And not only applicable to science.

Consider the confrontation between Aristotelian and Newtonian mechanics. For Aristotle, all motion required a moving force. This force could be, in the case of heavy bodies, the attraction towards its natural place, the centre of the earth, or a force applied to the body such as a push. Inertia was a state of rest and all motion had to be accounted for by reference to a force. This accords well with our everyday experience, an object does rest unless it is pushed and will stop moving if we cease pushing. Motion is a change of position. That is the language we speak. It was in giving an explanation of such observations that Aristotle was led to say;

> Everything that is in motion must be moved by something. For if it has not the source of its motion in itself it is evident that it is moved by something other than itself. (Aristotle, 1984, 241b 34-35)

At the centre of Aristotelian dynamics is the idea that bodies move against constant resistance, inertia is defined as a state of rest.

Our language contains within it the Aristotelian idea of inertia. The dictionary describes 'inert' as having no inherent power of action, motion, or resistance and 'inertia' as inactivity.

This is in contrast to Newtonian mechanics. The axioms or laws which appear in the *Principia* can be stated as:

- Every body continues in its state or rest, or of uniform motion in a straight line, unless it is compelled to change that state by force or forces acting upon it.
- The change of momentum of the body is proportional to the force acting upon it; and is made in the direction of the straight line in which the force is directed.
- To every action there is an equal and opposite reaction.

The basic principle of Newtonian mechanics lies in the first proposition. It is a law of inertia and is in direct contrast to that of Aristotle. To grasp Newtonian inertia we must imagine an idealised system. Aristotle demanded no such exercise of the imagination. Consequently, any student with sufficient command of the English language to use either 'inert' or 'inertia' is confronted with a problem when confronted by Newtonian mechanics for they will tend to speak an Aristotelian language.

Newton's second law is usually expressed as $F = m. \partial v/\partial t$ where F is force; m is mass; and $\partial v/\partial t$ the rate of change of velocity with respect to time, acceleration. A more common form of expression is $F = m.a$ where a is acceleration. Again, in contrast with Aristotle, we are asked to imagine an idealised system. Our everyday idea is of heavier bodies falling faster and there is never an absence of resistance. In modern terminology, it has been said that for Aristotle, $v = k\ F/R$ (where F is force; k, a constant; v, velocity; and R, the resistance) (Stinner , 1994, p. 78). This is an interesting attempt by a person speaking from within one tradition to translate the ideas of another tradition.

Newton's third law is again a case of an ideal system in which there not only can be action and reaction in contact but also at a distance whereas Aristotle's approach was to consider that there could not be action at a distance and that all action which produces a reaction must be by way of contact. Nor was the reaction considered equal and opposite but dependent upon the circumstances of occasion.

Aside from the detail of the laws, Newtonian mechanics demands that we see the world with which we are concerned in an idealised way. This was the approach used by Galileo in his work on the pendulum and it enabled mathematics to be applied to physical problems. Aristotle did not conceive of his mechanics as applying to an ideal system but to the real world. He is thus limited in the use to which he can put mathematics should he have wished to do so.

Clearly, the learning of Newtonian mechanics by students poses problems. The main problems are that students are asked to put to one side their everyday notions, that is, their Aristotelian notions, and enter via their imagination into a different way of seeing the physical world which they inhabit. It might be thought that this is merely the problem of learning another approach to put alongside their previous commonsensical approach. Students, as people, are accustomed by the time that they reach adolescence to wearing different hats for different occasions, this is just one more hat. And yet the anecdotal evidence and the evidence in the literature suggest that students find this very difficult at times.

Science teachers are confronted with a problem. The world that their students observe is an Aristotelian world. The problem for the teachers is to get the students to change their point of view and look at the world as an idealised system in order to comprehend modern physics. This problem has been commented on before (Ebison, 1993; Lombardi, 1999) with little more than injunctions for teachers to be aware of the problem.

The problem is not insoluble it is the problem of translation from one tradition to another. It is similar to the problem confronting language teachers when they attempt to get pupils to learn a second language. It is possible to learn a second language in the sense that it enables a person to get around a country and a culture without too much fuss but with little grasp or contact with that country and culture. If there is to be a worthwhile contact with that country and culture then the second language must become as a first language. Only then is it possible to understand and partake of the beliefs and attitudes of that culture. The same is true of different traditions. The scientist immersed in classical mechanics must learn the language of the Aristotelian tradition of mechanics if they are to understand the coherence and incoherence of that paradigm within a particular tradition. Only then will they be in position to ask whether or not the Aristotelian tradition brings with it insights which are superior or inferior to those from the tradition within which classical mechanics is embedded. In this way it is possible to compare and contrast paradigms within traditions rationally. Certainly there may be elements in a tradition and its paradigms which are incommensurable just as it is impossible to translate certain phrases from one language to another but, accepting this, it is still possible to make significant comparisons and assessments.

For comparison to be possible I suggest the following steps are necessary, though not necessarily in the order stated. The matter is not to be presumed to be linear:

- a translation of Aristotelian mechanics which is sufficient to express the claims made by the paradigm in terms that are meaningful in the statement of classical mechanics;
- an examination of the internal coherence of Aristotelian mechanics and classical mechanics;
- a statement of the claims which are incompatible between the two paradigms;
- an assessment of the extent to which classical mechanics removes the incoherence in Aristotelian mechanics;
- an assessment of the extent to which classical mechanics explains more of the physical world than Aristotelian mechanics.
- the explanations will be tested against evidence gathered from reality by means such as:
- direct observation;
- experimentation; and
- modelling.

Good science teachers do this. If the methodology were applied in the teaching of the sciences of other cultures within a Western culture then:

- the history of that science must be taught in a way which gives it meaningful expression;
- the coherence of the science must be examined;
- any incoherence, and there will be some, must be evaluated;
- was the incoherence bound to occur?
- a comparison must be made between the explanations offered by Western science and the other science to determine which provides the better explanation; and
- what counts as a good explanation in science must be taught.

However what has been ignored in the above is the problem of translation between radically different cultures. I shall return to this later.

It is evident that in attempting to translate, compare and evaluate the merits of different approaches the scientist and student has to stand back and ask questions of the two paradigms being compared. I use the notion of standing back deliberately for it seems to me that in this work the scientist and student is going beyond simply reasoning to a more fundamental evaluation. This I would term reflection and I shall return to the nature of reflection in this context later.

THEORY

Incommensurability raises the issue of the relationship between theory, evidence and experience. The first point to make is that observation, as a form of experience, cannot lead to theory without the observation itself being theory driven. A human being can observe apples falling from trees until eternity without making observations connected with scientific properties. The apple accelerates but what is meant by acceleration? Matthews (2000a, citing Lonergan, 1993) makes the point that the scientist means by acceleration '$\partial v/\partial t$'. To understand that mathematical formula is to understand acceleration, anything else is an approximation and, because of that, inaccurate and of limited use. Of course, this only applies if one is thinking within the terms of Newtonian mechanics. A point that neither Matthews nor Lonergan seem to realise. The implication of the theory and observation dilemma is that there must be a grasp of theory before useful scientific observation can be made. And the implication of that is that sufficient theory has to be taught to the learner before worthwhile observation can take place. There must be transmission of necessary concepts without which the prospective learner is wandering in a buzzing, booming confusion with little prospect of producing knowledge. Galileo was a genius present in a part of the world at a time when the circumstances in which he existed, and of which he knew, were ripe for the moves that he made. As Matthews points out, Galileo's circumstances included the influence of the music of the time and, in particular, polyphonic music which brought with it an emphasis on time as an measurable quantity (Matthews, 2000a, pp. 276-277). There were of course other factors in play. The point is that the social circumstances of the observer clearly affect the observation and what may be derived from it, but, these social circumstances are replete with theories that the observer brings to the experience and determine the initial nature of the observation. The process after that may well be further observation on the basis of new theory and so on in a chain of events that are not necessarily linear.

Observation may lead to the recognition of simple common conjunctions but that and its application do not make a science. Consider the making of poteen. Poteen has been made in the West of Ireland for centuries by a people who had no knowledge of the physics and chemistry of what they were doing. The process was complex. Was it put together by trial and error or was it informed and guided by some theory? Or did the theory come after to justify and hold together the process and allow it to be easily learnt by coming generations? Or was there no theory but a purpose driven activity? So, for instance, the point of making poteen was the whiskey. The problem was to separate the alcohol from the mash. It was observed that when the mash was heated the vapour it gave off condensed and tasted of alcohol. So the problem was partly solved – heat the mash and collect the condensation. But how to arrange the condensation? It was known that hot vapour condensed on cold objects. So arrange for the vapour from the mash to collect on a cold object. How to tell when all the alcohol has been collected? Alcohol burnt fiercely but water did not. Throw some of the liquid just collected on the fire and see if it burst into flame.

What we have here is an account of the growth of a complicated process starting with a simple observation that leads to an attempt to use that observation in the organised production of something that is desired. There is no knowledge of the chemistry and physics required. There is no explanation required of why it works as it does. There is no theory. There is just knowledge of constant conjunctions.

The next problem is how this process is to be passed on to future generations. A way of doing it is to incorporate the elements of the process into a story. The story is framed in such a way that it guides the distiller in what to do and provides a justification for the actions that he or she performs. The story does not have to be true, it just has to convey the process so that future generations can repeat it successfully. Stories gain accretions over time which may or may not be essential to the process. So typically future generations ask – 'why do you do that?'– answered by – 'I don't know it has just always been done like that'. Sometimes then the story gets changed.

The case of poteen, as described, is not a case of science but of technology based on the observation of the world and the application of some of those observations towards a specific end. The story that then might surround the making of poteen is a guide to the process not an explanation of it. A similar account can be given of making wine, bread, olive oil and many of the things that human beings have produced over the centuries. What is the difference between this and Western science?

The first and most obvious difference is the lack of theory underlying the process. But it might be said that the traditional science has a theory, it just happens to look like myth to us.

This leads to the notion that where the sole purpose is the making of poteen a suitable theory is one that does not disrupt the successful making of poteen and the passing on of this knowledge to future generations. It is irrelevant whether or not the theory in any realistic sense is explanatory of the world of our experience. There are a number of theories which would do the job for the poteen maker. For a number of limited purposes Ptolemaic astronomy was perfectly adequate. It is only when we seek to go beyond what is encompassed successfully by Ptolemaic astronomy that we need a new theory. It is only when we seek to go beyond what is encompassed successfully by the poteen theory that we need a new theory. But this still ties the need for a theory and the assessment of a theory to success in practical matters. It is a technological, utilitarian approach to the definition of a science and the construction of theory and, in this respect is not, I suggest, compatible with the nature of a practice whose aim is knowledge.

SCIENTIFIC VALUES

It might be argued that the practice of science results in two benefits, a benefit to the practice through the achievements of scientists, and a benefit to the society in which that practice takes place. I leave aside for the moment any benefit to the individual practitioner. The first benefit is judged by criteria internal to the practice of science and the second benefit is judged by criteria internal to the culture but external to the practice of science. The practice of science on this view progresses through assessment by criteria internal to the practice and independent of the culture, while the cultural benefit of the science is determined by reference to cultural criteria. This divorces the value of the practice from the cultural benefit that may flow from the practice. It leaves open the possibility that a piece of science may be of benefit to the practice of science but of no benefit, and possibly damaging, to the culture in which it appears. This form of argument raises its head in relation to the splitting of the atom and in relation to the attempt to teach Western European science to African villagers. It is also the form of argument that relies upon the traditional distinction between science and values. This dissociation of the internal benefits of science from the cultural benefits is found quite clearly stated in many curricula. It is emphasised that the value of science lies in its social benefits and is to be studied simply because it has social uses. This, of course, sits well with the utility of science in preparing students for a working life and with the notion that science is a social construction. If there are benefits seen in terms of the progress of the practice then these are of lesser importance, if indeed they are accorded any importance at all. It could well be that progress in the practice is defined in terms of social applicability and benefit.

There is a traditional distinction between science and values that asserts the distinctiveness of scientific values with the implication that they are valuable for themselves as part of science:

Given the emphasis on the social benefits of science, are these distinctive scientific values viewed as worthwhile because of their social benefits or because of their value to the practice of science, or both? If they are viewed as worthwhile because of their benefit to society then they can be seen as part of society's values and the distinction between science and values breaks down. If on the other hand they are seen as worthwhile because of their internal value then the emphasis on teaching science for its social benefits is undermined.

Are the distinctive, traditional scientific values culture specific or science specific? If culture specific then the claim is that in this particular culture, in the practice of science, there is value placed on, for instance, openness to new ideas, intellectual honesty and the critical evaluation of data and arguments. The implication being that other cultures do not or may not possess or practise these values. If such arrogance is ruled out then the assertion reduces to the belief that the practice of science in any culture has these distinctive values. The implication being that they are not practised elsewhere in society. But this too is surely indefensible. Is the assertion one that it is only in science that these values are learnt? Again this seems indefensible. I shall argue that the traditional view is indefensible and that there is no worthwhile distinction to be made between the values that are learnt and practised in science and the values that are learnt and practised in other practices and in society as a whole. Honesty, prudence, justice and fortitude do not exist solely in science but are necessary to the practice of science, to other practices, to the existence of society and the flourishing of human beings.

MODELS

The philosophic and scientific literature on models and experiments is very large. For the purposes of this book it is inappropriate to survey this literature. It is appropriate to take from the literature the arguments and ideas that are pertinent.

In Euclidean geometry, 'any two points lie on one and only one straight line'. This together with other axioms constitutes the formal system of Euclidian geometry. The structure of the geometry lies in the relationships between the axioms and their deductive consequences. A model in this context is any entity or set of entities which constitute an interpretation of the geometry, its theorems and axioms, and in which the theorems and axioms are true. Thus, geometrical diagrams of the various Euclidean theorems are entities which, viewed ideally, reproduce the relationships that hold between the axioms. The term 'model' in this context is one that occurs in the interpretation of a deductive system. It has been called 'formal' (Kaplan, 1964) and 'logical' (Hesse, 1967). This sense of 'model' occurs in science where mathematical theories are applied in scientific practice. The commonest use is probably in the application of probability to the behavioural sciences. In this case a mathematical theory containing probability axioms is used to generate a 'formal' or 'logical' model.

This use of the term 'model' is in contrast to its use in common parlance. A model here represents a likeness, a replica. Quite often it is also thought of as a miniature version of the thing modelled. Models are smaller and therefore it is possible to see far more, usually all, of the aspects of the model at once. Common parlance here supposes that the model is a physical system. So a model of a building project can be viewed with what might be termed 'a God's eye view', something which is normally impossible. However, to scale down is necessarily to lose something. An engineer viewing a model of a bridge accepts that when the bridge comes to be built it cannot be built with the materials and structural strength of the model, it would fall if it were. A model of the solar system is not a miniaturized copy but a model and therefore it lacks something of the real thing, the gravitational effect, the spatial void and so on. This is the case in planetariums. In the same way when we talk of persons modelling their behaviour on another we are not talking of copying that behaviour but adapting the perception of another's behaviour. A model is not a copy and it is a mistake to treat it so. It would be to treat the model as identical in all respects to the object modelled except in the matter of size. A model may however be a replica. In which case a model may be said to condense while resembling significantly that of which it is a model.

The terms 'resembling', 'like', 'analogous to', 'similar' and others all occur in talking about models. Hesse (1967, p. 355) makes a distinction between formal and material analogy.

- Formal analogy is that of 'structure or isomorphism between model and system'. There may also be a formal analogy between two materially dissimilar models of the same formal system.
- Material analogy is that between the object modelled and its replica. The replication may consist in some material identity of substance or shape but not of size.
- Models may be analogous both formally and materially.
- All analogies imply differences, negative analogy, and similarities, positive analogy.

The formal model mentioned above is one that has formal analogy. The model bridge of the engineer has material analogy.

Consider the use of a model in science. If we imagine a system under investigation then it is normal scientific practice to seek another well-established system which is sufficiently analogous and use that system to explain the system under investigation. Refracted light is produced by shining white light through a piece of glass. If we imagine light as a stream of particles or corpuscles which behave according to Newton's laws of motion then it is possible to explain the refraction. The rectilinear propagation of light becomes deducible from Newton's laws of motion provided we imagine light to be analogous to corpuscles. The phenomenon is explained by reference to a familiar theory. However, all analogies are both negative and positive. Hesse (1967, p.356) argues that when the negative analogy of a model is removed it may then properly be called a theoretical model in science. She uses the example of the billiard ball model of gases in which the size and colour of the balls is not intended to be part of the analogy. Size and colour are part of the negative analogy. A theoretical entity is left when the negative analogy is removed and this serves as the model.

There are dangers in the discard of the negative analogy. Kaplan (1999, p. 88) notes two dangers of using the term model, one the opposite of the other. Kaplan does not refer to Hesse and does not mention the notions of positive and negative analogy. However the dangers to which he points can be viewed as ways in which negative analogy can disrupt the successful use of a theoretical model:

- attaching significance to features of the model which are only part of the structure of the model and not of that which it is modelling; and
- omitting something from the construction of the model which is significant in that which is being modelled.

  To which might be added:
- inserting as part of the model something which does not exist in the object being modelled.

Removing the negative analogy produces a model in which, in addition to the positive analogy, there will be analogies that are not, at present, evident and part of the heuristic of the theoretical model lies in the investigation of this excess of analogy. Hesse (1967) terms this excess, 'surplus meaning'.

Kaplan approaches the concept of 'model' in science differently:

> Structure is the essential feature of a model. In an interpretation of the model, a correspondence is specified between the elements of the model and those of some situation, and between certain relations holding within each set of elements, so that when two elements of the model are in a certain relation the corresponding elements stand in the corresponding relations, and vice versa. A set of elements related in certain ways is a system; a structure is what is shared by corresponding systems (or it may be identified with the set of all possible systems corresponding to a given one and thus to each other). (Kaplan, 1999, p. 88)

Kaplan then goes on to describe a physical system which may be a model boat or airplane or shopping mall. Where the model is used in a way which preserves its structure he argues that the results show what would happen to other systems having the same structure. This is to make the point that models are about correspondence of structure and not about content, a point which he had made earlier (Kaplan, 1964). He continues:

> When the correspondences are only suggested rather than being explicitly defined, the symbolic system is an extended metaphor, intermediate between a metaphor and a model is an analogy, in which correspondences are explicit but inexact. All three have roles in the actual conduct of inquiry; the view that only models have a place in science makes both terms honorific.
>
> In another honorific usage "model" is a synonym for "theory" or even "hypothesis". The term is useful only when the symbolic system it refers to is significant as a structure - a system which allows for exact deductions and explicit correspondences. (Kaplan, 1999, p. 88)

While it is possible to see the distinctions that Kaplan is making they appear to be distinctions that can rarely serve a useful purpose in the practice of science. Hesse uses the term 'analogy' and this covers any system which bears a resemblance, is like or is a replica of another system. There does not appear to be a need to complicate matters further. Parsimony operates.

There is no doubt that theoretical models have a significant part to play in the practice of science. However, there has always been debate about the relationship between theory and theoretical models. Are models merely a device to aid in the understanding and production of theory or are they a necessary part of the theory? This is another rerun of the debate between positivist and realist theories of science. On the positivist side there is an insistence that all non-logical terms in a theory should have an explicitly empirical definition. This makes models dispensable when the theory is fully developed. On the realist side there is a view, which I accept, that theoretical models are an essential part of the theory (Campbell, 1957). The corpuscular model that enables an explanation of refracted light has its origin and interpretation in Newtonian mechanics. There the corpuscles do not have an explicitly empirical definition. Thus the positive analogy of the model becomes part of the explanation and part of a theory of reflection and refraction. This will be true wherever concepts in the positive analogy of the model do not have an explicitly empirical definition. For instance, any explanation that relies on particle mechanics, as the Gas Laws do, relies on a theoretical model which contains definitions that are not explicitly empirical.

In my view a model, in general, is an exemplification of the theory. A model is not a theory, it is an instance of a theory given particular form. It is always a case of resemblance not identity. It can, on occasion, usefully be thought of as a structural analogue of the theory. It serves a number of functions. It enables understanding of the theory and its application. It enables the scientist to modify, change, redirect and experiment in order to cast further light on theory and light on avenues of exploration. Importantly, a model is an example of the way the world looks according to that theory. A theoretical model is a particular case. A theoretical model, consisting of its positive analogy, is rich because it has been stripped of its negative analogy and joined with theory to explain something in the real world. What we have here are theoretical models derived from theory being used to produce further theory with the aim of explaining the world of our experience. Models are essential to the development, sustenance and defence of hard cores.

On any version of models it must be that models are in some fashion linked to theory. A model is an instantiation of a theory. There are also a number of possible models which may be used in relation to a single theory. It is clearly the case that science uses theoretical models and, I would argue, it is also the case that these theoretical models are derived from theory and are used to investigate theory. Unless theoretical systems such as Newtonian mechanics are considered models rather than an axiomatic system then this does not seem to correspond to the practice of science in the world as it is. To consider everything a model does two things:

- it may produce confusion by failing to make use of valid and useful distinctions that exist in the real world; and

– it leaves open the possibility that if science only deals with models then it never deals with the real world.

The latter emphasises the construction of models which may be personal or social and that it is these models that are the world for a person, society or culture. However, it must be remembered that all the caveats about the use of models set out above apply, if not, then there is a peculiar use of 'model' relying for its force on ordinary usage. For instance, I have remarked that the terms 'resembling', 'like', 'analogous to', 'similar' and others all occur in talking about models. It is a mistake to fail to acknowledge that those terms necessarily point to a 'something' other than the model against which the model is judged. Models may be constructed, pulled apart, reformed, modified but always with the view that doing this may throw light on a world other than the model. Eventually the touchstone is always a world which exists independently of our thinking of it.

EXPERIMENTS

> Basically, experimentation is a process of observation, to be carried out in a situation especially brought about for that purpose. The great astronomer Herschel characterized experiment as nothing other than "active observation", which is indeed what it is in astronomy. But no scientific observation, as we have seen, is wholly passive, how much the scientist intervenes before or during the process of observation is a matter of degree. (Kaplan, 1964, p. 144)

I am at odds with this view. An experiment is merely a device. It may be a physical device, it may be a thought, but it is a device. It is not observation for observation is what takes place when we observe the device. It may be said that to experiment is to observe, but that cannot be. The same device can be observed in different ways, different devices can be observed in the same way. If we lose the distinction between devices and observation then we lose what is useful. Devices are, of course, closely linked to models. We construct a model airplane and place it in a wind tunnel. We vary the speed and direction of the wind and collect the data by some means which has been built into the device. Perhaps we alter the inclination of the wings or the tailplane and collect more data. It might be said that our observation of the device in action leads us to change the inclination of the wings. But this conflates a number of things. The device produces data which are recorded. It is our observation of the record of the data produced by the device that may make us think that it might be interesting to alter the inclination of the wings and see what data are produced. But why should it do that? The observation is of the recorded data and that observation is guided by hypotheses derived from theory. It is the hypothesis or theory that provokes our interest in the recorded data. It is the hypothesis or theory which enables us to observe what has been recorded and reflect upon it. This is what happened when Einstein asked himself what data of the world would he record if he were to travel on a beam of light. This is to say that observing the results of an experiment is a matter of reflection. To observe is to do more than see. Science is an activity in which models play a part, in which experiments play a part, but science is essentially a reflective activity. And it is this that points up one of the main problems with many curricula. Typically such a curricula would also use the word 'reflect' throughout its description of what the students were supposed to do. The word 'reflection' and its synonyms appear scattered like confetti throughout curricula and reports by Departments of Education. Rarely, if ever, do the documents give a moment's thought to the meaning of the word. Yet it is not a word whose meaning is transparent. the curricula will talk of reflection and scientific values such as intellectual honesty and rigour without specifying any connections. I shall argue that values such as intellectual honesty and rigour, respect for evidence and valuing the views of others, are the products and the workings of reflection.

It is useful to consider the example of Galileo and pendulum motion. For thousands of years before Galileo human beings had observed pendulum motion. Objects had swayed backwards and forwards in a breeze. The rocking of a boat could set objects suspended in that boat into motion. Chandeliers had swayed above the heads of intelligent and perceptive human beings. As Matthews remarks (1994, p.111) people were searching for a reliable way to measure time because it so obviously held the answer to many of the problems that beset everyday life. Yet no one, apparently, saw isochronic motion in the pendulum before Galileo. The crucial change that Galileo made was to introduce idealised conditions for the swinging pendulum. If the swinging bob can be considered a point mass which is subject to no air resistance, no friction, no elasticity and so on, then, it is possible to introduce mathematics. Mathematics can work in an idealised system and the results can be referred back to reality (Gibbons, 1979). This brings out the notion that the experiment is not merely a device but a device conceived in a special way by the experimenter. All devices used by the experimenter are constrained. That is, the experimenter sees the device and uses it not as a realistic object, whether tangible or not, but as an idealised object. This affects the data that are produced and therefore the observations that can be made.

There are, however, the same dangers inherent in the use of an idealised system as those that were pointed out above in relation to models. The idealised system is essentially a model. As Matthews remarks:

> There is, however, a problem of idealization hiding fundamental mechanisms in the world. Keeping one's eye on the essential property is scientifically commendable, provided that it is the essential property, and that concentration on it does not blind one to other significant influences or properties. (Matthews, 2000a, p. 247)

The point here is that the experiment is something which the scientist constructs or imagines and, using what has been constructed, he or she attempts to investigate. The investigation will, hopefully, yield some worthwhile results. Experiments and the observations that are made of them are processes employed in the investigation of the relationship between hypotheses and theories and the real world.

What counts as worthwhile? The point of an experiment is to investigate an idealised or constrained system in the hope that it will throw some light on the nature of the real world. The biologist who injects material into a living cell is operating in a constrained and isolated world. The physicist who experiments with a system, whether tangible or not, is operating in a world that is imagined. Whatever results are achieved in experiments have to be checked against reality for the point of the experiment in the first place was to throw some light on that reality. This creates a problem. The data are the result of working with a system that is constrained and therefore the results from an unreal system. The problem arises in referring the data back to the real world. Consider the swinging bob again. A typical school experiment is to set up a swinging bob and measure the period of the swing (T) for various lengths (l) of the bob. A graph may then be plotted of T against l. If the experimenter is working from the formula $T = 2 \quad l/g$ where g is the gravitational constant, then, theoretically, the graph should be a straight line. It never is. With a nod to people such as Gauss the best straight line is drawn. The drawing of the straight line is an affirmation that, within the constraints of the experiment, there is correspondence between the formula and the world. That affirmation rests upon the accuracy of the experimental data. If it were impossible to obtain data that gave a spread of readings that justified the drawing of the best straight line then doubt is cast on the formula and the theory on which it rests. If more and better experimental technique leads to a greater approximation to a straight line then there is a justifiable assumption that there is correspondence with the real world. In practice this led to the formula for the compound pendulum. In practice, technology, when using scientific work, incorporates an estimate of the percentage error into its work and designs and builds accordingly. If it fails to do so the bridge falls.

There is one type of experiment that has played an important role in modern science and an examination of which brings out a great deal that is important in the role of experiments in science, the thought experiment. I have argued that any paradigm contains within it a fatal flaw, the effect of which will lead to modification or even discard of the paradigm. There was just such a flaw in the Newton's work on gravitation and the cosmos. Newton made two crucial assumptions, that both space and time were absolute. By the first he meant that space is to be thought of as an infinite linear plane. This is not our experience unless we confine our experience to our immediate surroundings. By the second he meant that time was the same from any vantage point. This was the belief that underpinned the creation of Greenwich Mean Time. Liebniz, Newton's contemporary, criticised both assumptions. He took both space and time to be relative. The criticism had little effect and the Newtonian assumptions were undisturbed for more than two hundred years. Michelson's work in the nineteenth century produced anomalies for the Newtonian paradigm but such was the weight of that paradigm that it was not doubted. Time was an absolute. However, there were also problems with the Newtonian interpretation of Maxwell's work on electromagnetism. Enter Einstein and perhaps the most famous thought experiment of the twentieth century. What would happen if I rode on a beam of light looking back at a clock? Time would stand still for me. At that point I am in a time and space capsule. In that capsule I can discover the same laws as a person standing beside the clock but the values that I will get are different. The value that remains constant is the speed of light. Time is relative. From this comes other thought experiments. Imagine the earth is a flat disc and imagine identical clocks set to the same time placed at the centre of the disc and on the turning rim. The clock at the centre will show more time passed than that on the rim.

This account fits with the account I have argued in relation to incommensurable paradigms and incommensurable cultural sciences. It is an account of rationality at work. The thought experiment is not confined to science but appears wherever rationality is at work. Searle's 'The Chinese Room' is one such example (Searle, 1999), as indeed are the replies of his many critics.

The experimental device is that conjured up by the imagination. It might be Einstein's ride on a beam of light, Newton's rotating bucket of water, Galileo's dialogue on falling bodies, Liebniz's imaginary experiment which attacked the Cartesian idea of momentum or Born's train of electrons riding on a crankshaft. In each case the experimenter has asked a question of an imaginary situation. These questions are often – if X is true then what would happen if …? or, what would the world be like if …? The first question is usually aimed at exploring the current theory and, perhaps, revealing its flaws. The second question is usually aimed at attempting to formulate a new theory. These are classic instances of human reflection at work. Note that the questions are not peculiar to science.

It is obvious that the human imagination can conjure up an infinite variety of scenarios and, thus, there must be some guidance placed on the production of thought experiments and their evaluation. This is derived from the theory that is being considered in the first place. However, this cannot rule out inconsequential thought experiments and their results. The results of the thought experiment must eventually be checked against the real world. In so far as this is possible this is a matter of checking the predictions from the experiment against actuality. So, Einstein's prediction from the General Theory of Relativity that the gravitational field of the sun would cause light to bend was tested by Eddington during the solar eclipse of 1919. The prediction was confirmed.

Clearly a thought experiment is an experiment with concepts. To ask students to imagine looking back at a clock as they rode on a beam of light and describe the clock after a period of time is to ask students to play with the concepts and that play will be revealing of the students' grasp of the concepts.

In science teaching the use of computer systems to replicate historical thought experiments is now possible. There is now the possibility of developing computer systems that enable the teacher and his or her students to develop and work with thought experiments. This is an important development since the computer comes to the aid of the reflective process.

Thought experiments are the reflective scientist at work. Thought experiments in the teaching of science are an aid to the understanding of concepts and to the development of reflective thinking.

CHAPTER 6

COMPARING WORLDS

The social acceptance of any idea comes about as a result of social negotiation. It is a political event. There will always be some who do not accept the idea but as long as sufficient in a society consider it acceptable then the idea will be part of the knowledge and reality of that society. Social negotiation is crucial to an understanding of constructivism and also a weak link.

There is a thesis that cultures develop their own view of the world. On this thesis what is meant by another culture is that it has an idiosyncratic view of the world. There is therefore the possibility of a person from culture A considering the world view of culture B and vice versa. A number of issues arise as a result. Is it possible for the person from culture A to comprehend the world view of culture B? Hempel remarked:

> If adherents of different paradigms did inhabit totally separate worlds, I feel tempted to ask, how can they ever have lunch together and discuss each other's views? Surely, there is a passageway connecting their worlds; indeed it seems that their worlds overlap to a considerable extent. (Hempel, 1996, pp. 129-130)

Kukla, in discussing this remark, comments:

> … Hempel's focus is on the fact that adherents of different paradigms who have lunch together are able to engage in fruitful discussions. My focus is on the even more remarkable fact *that they can have lunch together*. Regardless of how strained and unproductive their conversation might be, how does it happen that there's a common *venue* in which it can take place? Aren't the conversers supposed to be living in different worlds? (Kukla, 2000, p 102)

Incompatibility between cultural views, paradigms, worldviews must, necessarily, be characterised as incompatibility between this and that. To suppose that one cultural view cannot exist in the minds of members of another incompatible culture is nonsense. In order for there to be a meaningful X there must be a meaningful not-X. What is demanded of any culture or paradigm is that it is sufficiently characterised for it to be possible to identify the claims that are being made. What is necessary if any culture or paradigm is to be seen as incompatible with another is that its claims about a topic are given sufficient expression for it to be clear that the claims being made are incompatible with the claims of the other culture about the same topic. This is to dismiss radical incommensurability. Within each culture or paradigm judgements are made about that topic according to the standards and criteria prevalent in that culture. The judgements are consistent within the culture or paradigm but the judgements are incompatible between the cultures or paradigms. To see that they are incompatible with each other is to be able, at some level, to understand and apply the standards and criteria of the alien culture and to see that the application of these standards and criteria excludes the application of the non-alien criteria. Kuhn's extreme view of incommensurability is a mistake and, by the same token, the quoted remarks of Kukla and Hempel are to be seen as an attempt at a *reductio ad absurdum* argument which makes the reader question the relativist ideas without, of itself, destroying them. Except in extraordinary cases it will always be possible for the inhabitant of one culture to understand sufficient of another culture to be able to assess its internal consistency and to compare that culture with his or her own. The above has been termed 'methodological incommensurability' (Wong, 2001).

While radical incommensurability may be a mistake, is it possible to argue successfully that when there are competing cultures or paradigms each with its own background of beliefs and conceptual scheme, its own standards of justification, then, there is no rational possibility of deciding in favour of one paradigm rather than the other? Is it that all we can do is see the internal consistency of the paradigm whether it be our own or an alien example but, because the paradigm can only be judged by its internal consistency there is no possibility of a comparison for any comparison would depend upon giving more weight to one set of reasons than another? Because what counts as reasons are determined internally to the paradigm, therefore it is not possible to make a rational decision between competing paradigms. Consequently, it may be said there is no standard of rationality which operates independently of the paradigms. This is a straightforward account of why the matter is relative. That there may be overlap between cultures merely complicates the issues involved it does not alter the problem.

Epistemological relativism arises from the same source. It is asserted that where competing cultures or paradigms have their own standards and criteria, then, judgements as to the truth of a state of affairs will be made relative to those standards and criteria. As a result no culture or paradigm can be the basis for denying the truth claims of the others. Truth claims are internal for any attempt to assert that they are external is to assert that the truth claims of other cultures or paradigms are in some measure false and this option is ruled out by the presuppositions accompanying the paradigms or cultures.

If a search is made for arguments which support such relativism as a tenable approach to knowledge and reality then the pickings are lean. Kukla (2000) links relativism and constructivism and contrasts the relativism of constructivism and the realism of rationalism. While his arguments appear in the context of a critical analysis of social constructivism he explicitly links constructivism and relativism and his observations are relevant to the current argument. He picks out a standard argument in the debate between relativism and realism which relies on the empirical evidence of historical research supported by the type of anthropological research conducted by Latour and Woolgar (1979). This presents constructivism (relativism) and rationalism (realism) as competing empirical hypotheses. Scientific ideas may be explainable on both hypotheses in which case, Kukla argues, further empirical, historical research is irrelevant to deciding between the hypotheses. As Kukla comments:

> … if it's possible to devise both constructivist and rationalist explanations for every episode in scientific history, then constructivism and rationalism are *empirically equivalent theories*, like the standard Schrödinger–Heisenberg quantum mechanics and David Bohm's quantum theory … What this situation precipitates is *theoretical debate* wherein the conceptual properties of the theories – their simplicity, their prospects for unification with other favoured theories, etc. – are analyzed and evaluated. (Kukla, 2000, p. 39)

On this view even if one attempts to reconcile constructivist and rationalist approaches to science by asserting that scientific ideas are socially produced and verified or refuted by reference to an independent reality then this is simply to produce another empirically equivalent theory. Historical research does not win the case for either side. Stove (1998) is mistaken in this regard. The conclusion appears to be that meta-argument is necessary on both the relativist and the realist sides.

The argument from the explanation of the history of science is, it is claimed, the only argument in the literature for constructivism. All other arguments put forward are elaborations of defences against attacks. As Kukla remarks:

> The positive case for constructivism is that thin. But we should recall that the positive case for most other strong metaphysical theses is also thin. There are no arguments for idealism or materialism that rationally compel assent from any of its opponents. There are only manoeuvres for blocking the rational imperative to give one's thesis up. Purveyors of metaphysical hypotheses aspire only to a strong defence against objections. If constructivists can do that much, they'll be doing as well as anybody else. (Kukla, 2000, p. 45)

This accurately picks out the current malaise. The past is the scene of interminable argument with each side standing their ground and we have reached the present with no victor in sight and a retreat to the view that there can be no victor because everything is idiosyncratic.

It is clear that both constructivism and relativism, whether ontological or epistemological, are intimately connected. Neither ontological or epistemological relativism are new ideas and the arguments for and against have been with us at least since Plato. The basic argument for relativism is the diversity of human belief and opinion on the same topic or topics. This is an argument grounded in empirical observation. To argue that the diversity of opinion supports relativism is to assume two things:

- that it is possible to argue by induction from the particular to the universal; and
- that the non-relativist thesis relies upon everybody agreeing on everything.

It has been claimed that there are problems with the first to which I shall return and argue that induction can be reasonable. The second does not follow.

The arguments against relativism have been standard since the time of Plato. A standard argument against relativism claims to be not merely persuasive but irrefutable and to show that relativism is untenable because it is incoherent. The argument runs: if relativism is true, then, the belief in relativism is held to be warranted, but the thesis is that nothing is warranted non-relatively, therefore relativism is unwarranted. That simple argument has been going the rounds since Plato. If the relativist attempts to respond by arguing that relativism is only relatively warranted then that assertion falls to the same non-relativist argument. The relativist is caught in infinite regress in the attempt to sustain his or her position.

Infinite regress has usually been held to be destructive of any argument or statement which falls prey to it. Hence the phrase 'vicious regress'. There is therefore a problem to which the relativists must produce a reply. There have been a number of attempts. One such attempt is based on the argument that in the practical world there is a pragmatic end to the regress (Ashmore, 1989, cited by Kukla, 2000). It has been argued that there are benign regresses and vicious regresses (Kukla, 2000). The distinction between the two appears to lie in whether or not there is a need to pursue the regress in order for the initial stage to exist meaningfully. So, if fact A is relative and this fact itself is relative and so on *ad infinitum* then, in order for fact A to exist meaningfully we are involved in a regress to which we cannot call a halt. On the other hand if I believe that all beliefs have a cause, then, while my holding belief A has a cause which is itself a belief and so on *ad infinitum*, there is no necessity for me to pursue the regress in order for me to hold meaningfully the initial belief.

Clearly the relativist position is one which leads to a regress. But is this necessarily fatal as countless philosophers have supposed? The argument that infinite regress is fatal presupposes that it is impossible to complete or conclude an infinite number of operations in finite time. But this in turn presupposes that each individual operation occupies a distinct time interval and this presupposition is an assumption for which justification is conspicuously lacking (Kukla, 2000).

Let us return to the distinction between benign and vicious regresses and alter the belief example. Suppose that for any act X in a game, A will only do X if she believes that B will do X and that A knows that B will only do X if he believes that A will do X. This, which Kukla describes as a sort of social version of an Eliatic paradox leads to infinite regress. The game can be completed by both A and B coming to an understanding and making the move X.

> But then, since the game takes only a finite amount of time, how can such an understanding ever be established? The answer, once again, can only be that the infinitely many parts into which the understanding is decomposed must not correspond to real-time steps … Why can't constructivists say the same about the infinitely many facts that have to be constructed in order to constitute any single fact? This is not a rhetorical question. There may be a good answer to it; but realists haven't given it yet. (Kukla, 2000, p. 78)

There may be an answer or two. The first point to note is that if there is infinite regress whether conceived of as occurring in real time intervals or not, then, there is no possibility of stepping outside the regress and considering its elements and evaluating them. Consider the following position of someone holding that all facts are constructed internal to and relative to a particular world view:

- All facts are constructed.
- The fact that this fact is constructed is constructed.

Now make this a personal statement:

- I am constructing this fact.
- It is a fact that I am constructing this fact.
- It is a fact that I am constructing the fact that I am constructing this fact.
- I am reflecting on the fact that I am constructing the fact that I am constructing this fact.

Now make it a more personal statement:

- I am constructing this fact.
- I am constructing the fact that I am constructing this fact.
- I am reflecting on the fact that I am constructing the fact that I am constructing this fact.

The sentences which include 'reflecting' cannot be meaningfully asserted if the premise is that all facts are constructed. The reason is that the attempt to introduce reflection breaks the chain of regress. The notion of stepping outside the regress and considering its elements and evaluating them is to reflect. Reflection, whatever else it may be, is a process in which we step outside our actions and our thoughts. It appears that the relativist above denies the possibility of reflection. The situation is analogous to an argument against Skinnerian behaviourism – if all human beings are subject totally to stimulus-response mechanisms, then it is not possible to step outside the system and appraise it, that is, reflect on it. This becomes a problem for Skinnerians for in putting forward their thesis they are inviting reflection. The argument must be distinguished from the psychological situation in which a scientist might find himself or herself when they are unable to think outside a paradigm. The one situation is a necessary conclusion, the other an inability. How can sense be made of the notion that relativism rules out reflection?

Human beings have language. There are a great variety of speech acts in any human language. We request, assert, question, greet, advise, warn and conduct a myriad of social interactions through the medium of language. So for instance, warning is concerned with a future event and the speaker has reason to believe that the event will not benefit the hearer. The speaker, under the impression that in the normal course of events the hearer may not know of the danger gives, by speech, an undertaking to the hearer that the event will not benefit the hearer (Searle, 1969, pp. 66-67). Searle comments that 'most warnings are probably hypothetical'. Searle's comment could be made stronger. For instance, a warning is advice about something which is non-beneficial. Advice can be about that which is beneficial or non-beneficial (I ignore the technical use in which only information is conveyed). Advice is always hypothetical and warnings are subsets of advice. This could lead to the position that non-human animals, who evidently can warn each other, can deal with hypotheticals. If non-human animals can choose then this would seem correct.

This outline of the speech act 'to warn' illustrates one use of language. In many respects it is a use we share with our animal kin. This, of course, is to take the view that non-human animals can have reasons for action. MacIntyre (1999) argues this point. It is a view that I accept. But human beings have developed another use of language which takes them beyond the use to which apes, dolphins and the like develop and use language. We have developed the type of language that enables us to stand back and evaluate what we have said or what we have done or what we intend to do. MacIntyre says of the language required for this use:

> What is needed in addition is the ability to construct sentences that contain as constituents either the sentences used to express the judgment about which the agent is reflecting or references to those sentences. Where my reason for acting is or has been of the form 'Doing x will enable me to achieve y' where 'y' stands for some good, reflection on this reason will require me to ask 'In this type of situation do I have a better reason for acting than that doing x will enable me to achieve y?' I will have to compare *this* as a reason for acting with alternative reasons for alternative actions. (MacIntyre, 1999, p. 54)

As MacIntyre rightly points out, this acquisition of the language of reflection is a pre-condition of rationality. Rationality is only intelligible in the context of the ability to reflect and consider whether or not this reason or that is appropriate, is relevant. As MacIntyre (1999, ch. 5) argues, Heidegger took the view that non-human animals responded to their environment but were trapped within it. Non-human animals could not take the step of standing back from that environment and considering it. McDowell (1994) embraces this view when he argues that non-human animals are driven and shaped by their biology and their environment. As a result both Heidegger and McDowell are committed to the view that non-human animals cannot have reasons for action. As a result McDowell is forced to identify reason as a defining characteristic of human animals. This is a mistake. Reflection is a defining characteristic though it may not be possible to draw such a sharp line.

If relativism is trapped in an infinite regress then relativism is a view that human animals are trapped inside their environment and are driven and shaped by it. Human animals are taken to be subject to external and internal forces.

A person trapped in such an infinite regress cannot have the ability to reflect. He or she is committed to a denial of the ability to step outside the regress and yet he or she must express their ideas in the language specifically developed to enable reflection and reflection takes place outside the regress. This is an uncomfortable situation!

The lack of comfort is more so if language and the future are considered. Talk of the future presupposes a use of the imagination to conjure up a variety of possibilities in which I might find myself. I am distant from them but can evaluate them, that is, I can reflect on them. Talk of the future brings into play concepts such as 'hope', 'despair', 'prudence' all of which require the ability to reflect and which, as far as we know, do not apply to animals other than ourselves. Can a dolphin hope? Can a relativist hope?

MacIntyre (1999) has produced an extended argument about human flourishing from the course of which it is possible to extract the thesis that the capacity which is distinctively human is expressed in at least four things:

- the ability to distance ourselves from our beliefs, decisions and actions both in the past and the present;
- the ability to evaluate our beliefs, decisions and actions;
- the ability to imagine and attempt to choose, which presupposes evaluation, our future; and
- the ability to imagine, which presupposes evaluation, our past.

And all four depend upon a particular form of complex language being acquired by the human child. All four presuppose an independence. All four presuppose that it is possible to step outside the regress thus destroying it. That is, it is self-contradictory to assert that all facts are relative and make that assertion in the language of human rationality and invite reflective humans to accept the soundness of what is asserted. One answer to Kukla's puzzle is that it matters not whether the regress involves separate intervals of time or not, for it is the language in which the regress is stated that destroys the relativist thesis by making any statement of the regress incompatible with it. The concept of reflection will receive further treatment later.

Of course what has just been argued presupposes that the principle of non-contradiction is accepted. The solution for the relativist is the repudiation of human rationality and the principle of non-contradiction. A number have done so. But this raises another problem. If the word 'white' is to have determinate meaning then there must be something which is 'not-white' otherwise white becomes the universal predicate which fails to pick out anything. 'White' has no meaning where there is not a 'not-white'. This is the principle of non-contradiction at work in language. To deny the principle of non-contradiction is to deny that which makes language meaningful. This can be put another way. The principle of non-contradiction asserts that X and not-X cannot both be held to be the case at one and the same time. If the principle is discarded then it presumably becomes possible to hold both X and $(\text{not-X})^n$. How can sense be made of this?

As always there is a response. If there is incoherence is not this the same with any paradigm? All paradigms have within them the seeds of incoherence whether it be the Newtonian paradigm, the Einsteinian paradigm or another. That is the nature of our inquiry into the world of our existence. Incoherence can be admitted and accepted as the norm. For this reason it is never enough to show that incoherence exists. It must also be shown that there exists a better paradigm, a better approach.

THE ENLIGHTENMENT

Before attempting to build the positive case it is important and necessary to suggest the reasons for the present flaws. When a point of view, an approach, a paradigm comes into existence it contains within it the incoherence which will eventually lead to its overthrow or modification. Therefore the argument must start with history.

The arguments both for and against relativism have, as I have remarked, been around for many centuries. Why is it that neither side can convince the other? Is it that neither the relativist nor the realist have convincing arguments? Is it the case that both relativist and realist are irrational? Margolis (1991; cited in Kukla, 2000) considers this a strange state of affairs which makes him suspicious. How is it that neither side can convince the other of the worth of its arguments? Who is neurotic? Who is irrational? I suggest that it is not that there is an unbroken level of argument between the relativist and the realist but that the argument rises in new forms with those revolutions of human thought that divide and excite. One such revolution occurred with the Enlightenment.

MacIntyre is of the view that what I have called 'ontological relativism' and 'epistemological relativism' in the modern world arise from Enlightenment and post-Enlightenment thought:

> While the thinkers of the Enlightenment insisted upon a particular type of view of truth and rationality, one in which truth is guaranteed by rational method and rational method appeals to principles undeniable by any fully reflective rational person, the protagonists of post-Enlightenment relativism and perspectivism claim that if the Enlightenment conceptions of truth and rationality cannot be sustained, theirs is the only possible alternative. (MacIntyre, 1988, p. 353)

For MacIntyre:

> The relativist challenge rests upon denial that rational debate between and rational choice among rival traditions is possible; the perspectivist challenge puts in question the possibility of making truth-claims from within any one tradition. For if there is a multiplicity of rival traditions, each with its own characteristic modes of rational justification internal to it, then, that very fact entails that no one tradition can offer those outside it good reasons for excluding the theses of its rivals. (MacIntyre, 1988, p. 352)

and the perspectivist solution to this quandary:

> … is to withdraw the ascription of truth and falsity, at least in the sense in which ‘true’ and ‘false’ have been understood so far within the practice of such traditions, both from individual theses and from the bodies of systematic belief of which such theses are constitutive parts. Instead of interpreting rival traditions as mutually exclusive and incompatible ways of understanding one and the same world, one and the same subject matter, let us understand them instead as providing very different, complementary perspectives for envisaging the realities about which they speak to us. (MacIntyre, 1988, p. 352)

The argument is that the Enlightenment was bound to lead to some relativist position. This is not to deny the benefits of the Enlightenment just as the identification of the flaws of Newtonian mechanics does not deny the existence of the benefits of that paradigm.

The overthrow of Aristotelian physics by Galileo was followed by the work of Newton and brought in its train an interest in the nature of science. Such was the predictive success of the new science that the thinkers of the Enlightenment came to believe that it represented knowledge and that other spheres of thought were either mere belief or could be subsumed under science. As Oakeshott said of later philosophers:

> There are philosophers who assure us that all human utterance is in one mode. They recognize a certain variety of expression, they are able to distinguish different tones of utterance, but they hear only one authentic voice … namely, the voice of argumentative discourse, the voice of ‘science’, and all others are acknowledged merely in respect of their aptitude to imitate this voice. (Oakeshott, 1962, p. 197)

This is the point which, among others, Marcel (1951), Oakeshott (1962), Habermas (1972) and MacIntyre (1981) have argued. All legitimate knowledge is equated with scientific knowledge, other claims to knowledge are dismissed. We are living in an age in which the achievements of science have seduced us into devaluing the intellectual achievements of those who come to know other than through science.

Truth had been achieved and was no longer the goal or *telos* of the inquiry but was realised throughout the course of the inquiry. Newton had put the heavens in their place. There were problems with this view of science. Hume attempted to found a scientific study of human nature and to incorporate ethics within it. In the course of his work he came to believe that one must distinguish between that which is normative and that which is natural such that it is not possible to derive evaluative conclusions from empirical premises or to define normative terms in empirical terms. It is not possible to get an ‘ought’ from an ‘is’. Since Hume’s time this particular dictum has occupied a major and influential place in philosophy. It is looked upon as a watershed in human thinking by some (Benn and Peters, 1959, p.16), others see the difficulty that modern philosophers have in understanding the approach prior to that watershed:

> It is difficult for a modern philosopher not to either misconstrue or wholly condemn Aristotle's ethical thought. This is because Aristotle begins by implicitly rejecting the basic modern contention that value judgments either are not judgments of fact at all or, if they are, are judgments of a kind of act altogether different from the facts involved in judgments about the physical world. (Kerferd, 1966, p. 161)

The problem for Hume and other thinkers then became the problem of exploring the nature of science and treating the problems of moral philosophy as distinct. Reasoning, for Hume, was concerned solely with inferences from matters of empirical fact or in the relations between propositions as in mathematics. Reasoning had no place in moral thinking or normative thought in general. To have thought otherwise would be to allow reasoning and deductive thought into the normative realm and thus the possibility of deriving evaluative judgements from empirical premises. Moral thought, for Hume, is expressed in propositions which convey approbation or disapprobation, that is, a feeling or sentiment which is particular to the person expressing the proposition.

In empirical matters Hume argued that all inferences from the existence of one object to that of another are based on the notion of cause and effect. He then defined cause and effect as simply signifying that two observed objects were in constant conjunction in time and space. Experience shows that an object released above the earth falls towards it. There is a constant conjunction, the first of which is called a cause and the second of which is called an effect. What we have is simply a constant association of ideas. Experience has continually shown us that A invariably follows B and so we come to firmly associate A with B. As a result Hume was led to conclude that reasoning formed no part in our recognition of cause and effect relations but was the result of past experience, of custom and of habit.

This account of physical phenomena is allied with the argument that we perceive the external world as a series of sensations, feelings and images. If we attempt to go from these perceptions to knowledge of the physical world then an inference is called for. But Hume denies that such an inference is possible for inductive logic is not possible. Hume argued that it was logically impossible to predict future behaviour on the basis of past, observed events. That the sun has risen many times in the past does not enable the logical conclusion to be drawn that it will rise tomorrow. There is no such thing as inductive logic. Independently of this notion Hume was a sceptic with regard to inductive reasoning.

> The premise of an inductive argument, Hume says, is no reason to believe the conclusion of it; a proposition about the observed is never a reason, however slight, to believe a contingent proposition about the unobserved .... This is his famous scepticism about induction. (Stove, 1998, p. 113)

Yet it is clear that in science such reasoning seems to be used frequently as indeed, noted Hume, it is in everyday life. Hume's response to the problem was to argue that scientific theories and laws, while, apparently, informative, can never be known to be true on the basis of reason and experience. He then asserted that the sciences were grounded in habit and instinct.

It has been said (Stove, 1998) that the influence of Hume on twentieth century philosophy of science and philosophy has been incalculable. The basis from which he proceeds, the impossibility of inductive logic or reason and that no observation can be a reason for a proposition about the unobserved, leads inevitably to the belief in the fallibility of any scientific theory. Once these ideas are accepted then, necessarily, it must follow that there is an acceptance of the impossibility of positive proof of a scientific theory. The result for someone like Popper, who accepted Hume's thesis, is that scientific theories can only be falsified. The result for someone like Kuhn is that paradigms arise within the community of scientists that serve as explanations of the physical world but do not have any necessary connection with the truth of the reality of that physical world.

Hume died in 1776 yet his influence on the philosophy of science is most marked in the twentieth century. Newton had died in 1727 having ushered in a remarkable analysis of the physical world that seemed to his contemporaries and those that followed him to provide the means by which the physical world could be understood, explained and confidently predicted. Newtonian mechanics appeared to supply the complete answer to Hume's scepticism. The philosophy of Kant can be seen as providing a supportive backdrop to Newtonian physics and the counter to the Humean thesis.

Kant argued that there are certain propositions which any rational individual must accept on pain of lapsing into incoherence. Kant was seeking criteria independent of any particular set of beliefs or theories, independent of any individual or society, to which an individual might refer in attempting to decide between competing beliefs or theories, or in attempting to evaluate a theory or belief. The basic Kantian question is always 'what is presupposed?' Kant was greatly influenced by two developments, Newtonian mechanics and the French Revolution. Using the first, the empirical world in which human beings live is understandable and there is a natural order of things. Using the second, social laws can be seen as made by humans. In the scientific sphere humanity creates laws by the use of reason but the truth or falsity of these laws does not depend upon humanity but upon the empirical facts. In the social sphere humanity can create laws but the truth or falsity of these laws depends upon rationality not upon empirical facts. There is always a gap between empirical facts and normative rules. Kant's approach was an attempt to overcome the problem posed by Hume's distinction between 'is' and 'ought' and Hume's denial of inductive logic and scepticism about inductive argument.

In the meantime there were enormous advances in the physical and biological sciences. The predictions that were successfully made using Newtonian mechanics created a great impression and the sciences of chemistry and biology began to follow suit. For instance, in the 1860s it was well known among chemists that the various elements exhibited family likenesses. The alkali metals were a family that exhibited a gradation in properties. The halogens were another family. The puzzle was how to explain the family likenesses. Mendeleev solved the puzzle in finding a mathematical key which enabled him to produce the Periodic Table. Arranged in the order of atomic weights with seven elements to a vertical column the families appear in the horizontal rows. However at that time, 1871, it was assumed that not all the elements were known so gaps were bound to appear. Mendeleev's solution was to fill in the gaps with what, mathematically, should be the next atomic weight. On the basis of his solution he predicted the existence of germanium and also predicted that it would have properties intermediate between silicon and tin. Twenty years later he was proved correct. But why the pattern? Six years after the discovery of germanium Thomson discovered the electron and so the atom could be seen to have constituent parts and the pattern started to become explicable. And this work might be seen as the start of the downfall of Newtonian mechanics. Almost contemporaneous with the work of Thomson came the work of Michelson and Morley who produced evidence that light did not behave as Newtonian mechanics had suggested it would. Shortly afterwards Einstein produced a revolutionary account of the fundamentals of physics. As scientists followed up Einstein's work it became clear that Newtonian mechanics could no longer give a satisfactory account of the physical world even if it still provided a useful tool in the ordinary world.

With the fall of Newtonian mechanics the belief that science could be the source of knowledge and immutable truth came under attack. Kant's reliance on Newtonian mechanics caused doubt about his philosophical theories. Hume's attack on inductive logic and rational ethics became a focus for thinkers resulting in the emotivist theory of ethics and positivism in science. Though it passed unnoticed what Hume had observed – people do not behave as sceptics in relation to inductive thinking.

At the beginning of the twentieth century there were two major attitudes which directed both science and moral philosophy. The first attitude was epitomised by the Kantian attempt to pick out the criteria to which any rational person must assent. These criteria were to be seen as independent of cultural differences and were the means by which it was possible to judge whether or not this or that approach, scheme or theory was the better because it was the more rational. Applying these criteria science could be a source of knowledge and truth. In moral philosophy and in the sphere of human action it was possible to compare and judge cultural moralities. There were no incommensurable theories, beliefs or paradigms. A significant example of this approach in a practical sphere was the work of Peters from 1959 onwards. For Peters moral thinking was based on formal principles such as 'make no distinctions in advance of a relevant difference' or 'have respect for persons' (Peters, 1966). These principles were looked upon as providing the context within which human beings could make moral decisions. They were, it was argued, principles to which any rational being must assent. As they were formal they contained no substantive content and, to that extent, were immune to substantive argument. Working from these principles Peters attempted to draw out what should be meant by authority, equality and other moral concepts when applied to education. The problem with this Kantian approach was the problem with all Kantian approaches, they were little more than injunctions to be rational.

They were divorced of social content, they had no substantive past or future, and as a guide to making the practical decisions required of moral agents were of little use. They were in a literal sense, unreal. They became mired in an endless debate about the precise analysis and its results.

The second attitude derived from Hume and the shock following the fall of the Newtonian world. All scientific theories were fallible and there was no possibility of proof. Morality was a matter of personal feelings and emotions which were, to some extent, influenced and developed by the culture in which the individual came into being, grew and existed. There was no possibility of cross-cultural judgements. If we wished to influence another person to act in a certain way then the problem was not one of convincing that person that you were right on the basis of the reasons that you could offer but of convincing the person by influencing his or her feelings or emotions. Manipulation became the basis of social activity. The Humean approach distinguished means and ends. It was possible to reason about means but ends were no more than an expression of the passions. Society was sustained by habit and convention and the moral decisions were decisions about ways and means not about moral ends.

It was an inevitable progression from this point of view to the view that the ends of morality were matters of individual preference. This is a common contemporary view in the West. Modern society is concerned with the ways and means, the ends are matters for individual preference, wish or desire. Hence the emphasis on money, profit, economics and power for these are universal means. Hence also the modern emphasis on manipulating people to obtain what is desired for there is no point in reasoning about ends. Society becomes a market place in which each individual or group is free within the rules of the market to seek what they desire. They do so by the manipulation of other individuals or groups. There is, of course, a reaction to this. On the one hand we had a British Prime Minister claiming that there was no such thing as society only individuals, the government and the market place and let the devil take the hindmost. On the other hand we have the attempt, often in the communities of the poor, to engender a sustainable community. Birch (1993) drawing on the work of Daly and Cobb (1989) cited the example of the Indian State of Kerala where the labouring poor and peasants benefited from the supportive environments of small communities underpinned by a State which saw the primary importance of the needs and welfare of all. This resulted in a State which, in respect of the quality of life of its inhabitants, was and is far in advance of the rest of India. The overwhelming rise of competitive individualism in Western society is an important aspect of the Enlightenment to which I shall return in the context of the East Asian challenge to it.

The adoption of the Humean approach to induction must be seen as containing two elements. First, there is the thesis that induction is an invalid attempt at logical argument. Second, and historically as important, is the thesis that the premise of an inductive argument is no reason for belief in its conclusion. Logical positivists accept the first proposition but not the second. Stove (1998) argues that Popper, Lakatos, Kuhn and Feyerabend accept both.

The overall result is the belief that argument must take the form of deduction and that scientific laws and theories are the premises for deductive arguments. Science becomes a nomological-deductive system guided by falsifiable premises. For Popper the mark of a scientific theory is that it is falsifiable, to the extent that if the theory is not, then it is not a scientific theory. A scientific theory is better than another if it is more disprovable. However, as Lakatos (1978a) pointed out, it is inherently impossible to falsify the premises because it is always possible to produce assumptions which save the situation. This, in itself, leads to a further complication. If it is always possible to produce propositions which save the day then no scientific theory can be incompatible with any other scientific theory.

At this point there is a situation which defies commonsense. Every morning I am able to observe the sun appearing over the eastern horizon. Am I being told that these observations over decades amount to nothing and that I am not entitled to believe that it will appear over the eastern horizon tomorrow? Yet that is the implication of Humean philosophy and leads to an irrational relativity.

> While Hume fails to solve the puzzle he raises, he is right to stress that we do not seem to act as if the puzzle were a problem. It seems that no one really believes that scepticism about induction is reasonable. To claim there is a problem of induction is inconsistent with the commitments we display through our actions. (Couvalis, 1997, p. 43)

True, but this may simply be a remark about human psychology. The approaches of Hume and Kant are clearly incompatible. The situation is exacerbated by the inability of the Kantian approach to produce anything which can be accepted as independent criteria capable of doing the job. As a result both in science and morality there has developed a belief that all we can do is continue with our individual and cultural endeavours and accept the fact that knowledge and truth are not possible in either sphere.

The message is that despite over 200 years of endeavour all approaches to the nature of science and the nature of moral thought have failed. The last century has seen mutually incompatible responses and argument to which there is no end and no result. Witness the science wars! The result is a belief that reason cannot solve the problem. A belief, it must be noted, that does not follow from the present inability to find a solution, if one accepts that induction is logically flawed. And if it were left at that then what has been written to this point would be merely adding to the stockpile.

MANIPULATION AND THE ENLIGHTENMENT

The Enlightenment, as has been argued above, produced in its train a collapse of justified and justifiable moral values. With Hume came the distinction between 'is' and 'ought' that divided moral thought and values from that which can be known. It was seen as a distinction between the rational and the non-rational. The failure of Kant and the utilitarians to produce anything which could put moral thought on a rational basis resulted eventually, with the help of the collapse of Newtonian mechanics and scientific certainty, in what is called 'emotivism'. Emotivism is a theory of ethics which asserts that all ethical assertions are assertions of personal preference and it is not possible to argue successfully that this or that preference is morally wrong for there is no principle against which such preferences may be judged. It is quite evident that a significant part of Western society view morality as an expression of personal preference and not rationally debatable. In such a situation there can be no rational attempt to convince other people that their moral view is incorrect and that they should accept as justified another view or society's view. The problem is a matter of non-rational persuasion not of argument. The problem is one of manipulating the other person to accept a particular point of view.

Kant made a distinction between manipulation and non-manipulation by stating that it was a moral imperative that a moral agent should never treat another solely as a means but always as an end. To treat someone else as a means is to treat them as an object to be manipulated to the end of another's purposes. To treat someone as an end in themselves is to treat them as a moral agent who may not be manipulated on pain of infringing the moral imperative. Modern culture in much of the West sees no Kantian distinction for there is no possibility of rational decision on moral values. Of necessity, manipulation, treating others as means to another's purposes, is the method of bringing another around to your view or society's view.

In such a culture management flourishes. Management is about seeking and implementing effective means to ends which are neither examined nor debated. The modern manager is the manipulator whose measure of success is the achievement of given ends. So managers are employed who do not necessarily have any views or deep understanding of the ends that they serve. Their work lies in the effective realisation of those ends.

In such a culture, education is about fitting the young to be effective manipulators and to be effectively manipulated. One of the goals of effective management is the creation of a skilled, flexible workforce for, with such a workforce, management can be successful. Key competences have to be identified and taught for by the schools. Key competencies are those competencies which the young must acquire in order to be effectively managed and, for some of them, to be effective managers. The unquestioned end is economic success defined in particular unexamined terms. The process of aligning the school curriculum to this culture was developed during the 1980s and 1990s in the UK and the USA. Other countries followed suit. An example of the reach of the ideas is given by its effect in Australia and on the small population of one Australian State, South Australia.

The Finn Report (Australian Education Council, 1991) considered the participation by young people in education and training that was available after the years of compulsory schooling and it introduced the notion of key competencies. This was followed by the Mayer Report (Australian Education Council, 1992) which addressed the issue of how to put general education to work. The Mayer Report is significant because its focus was clearly on the employment of the young and the notion of key competencies. In the years following the Mayer Report a curriculum for all South Australian schools was developed and, as the curriculum was reaching its final stages, a number of other reports appeared. The Allen Report (Allen Consulting Group, 1999) focused on the needs of Australian industry and, in particular, what was needed in order to ensure the competitiveness of that industry. Following the Allen Report there appeared a Report on the skills that were essential for employability (Australian Chamber of Commerce and Industry and Business Council of Australia, 2002). Throughout the decade the emphasis was on employability and the function of general education in the service of employability. The notion of key competencies was introduced at the beginning of the decade and was a central concept for all that followed. Key competencies were central to the new South Australian curriculum. Australia was not alone in focusing on key competencies. As the Mayer Report noted, the United Kingdom, New Zealand and the United States of America had all developed the same ideas. The question of ends, values and morality is not on the agenda for such things are a matter of personal preference. As the Mayer Report (1992) notes, such matters cannot be assessed credibly and are therefore not to be considered. They have no truth value because there is no possibility of credible assessment.

The management culture makes effective manipulation the controlling focus of the life of society. The therapist culture performs the same denial of the Kantian distinction in the private life of individuals. The correlate of economic success in the life of the individual is the well-adjusted individual. The end of normal adjustment is not debatable, nor is what is to count as normal or adjustment. The problem is to devise effective means to produce the well-adjusted individual. So the agenda is to produce a measure of self-worth, a measure of self-identity and so on. MacIntyre makes the point succinctly:

> The manager represents in his *character* the obliteration of the distinction between manipulative and nonmanipulative social relations; the therapist represents the same obliteration in the sphere of personal life. The manager treats ends as given, as outside his scope; his concern is with technique, with effectiveness in transforming raw materials into final products, unskilled labor into skilled labor, investment into profits. The therapist also treats ends as given, as outside his scope; his concern also is with technique, with effectiveness in transforming neurotic symptoms into directed energy, maladjusted individuals into well-adjusted ones. Neither manager nor therapist, in their roles as manager and therapist, do or are able to engage in moral debate. (MacIntyre, 1981, p. 30)

Then the question is if these ideas are applied to the schools what has happened to education?

CHAPTER 8

INDUCTION AND NATURAL KINDS

It is time to take stock and ask the question – is there something about the starting point which shows how the Enlightenment problem occurred? A starting point in science was with Galileo followed by the work of Newton. It was in the aftermath of this scientific revolution that Hume started to work. But the starting point cannot be seen as a starting point except in terms of that from which it departed and that was the medieval version of Aristotelian physics. If it is accepted that Galileo overthrew Aristotelian physics then is there anything about Aristotelian physics or the Aristotelian view of the world which should not have been discarded and, because it was discarded, produced the confusion of the next centuries? MacIntyre puts it slightly differently:

> It is therefore worth asking whether the Enlightenment may not have contributed to our present condition in a second way, not only by what its achievements in propagating its distinctive doctrines led to, but also by what it succeeded in excluding from view. Is there some mode of understanding which could find no place in the Enlightenment's vision of the world by means of which the conceptual and theoretical resources can be provided for reuniting conviction concerning such matters as justice on the one hand and rational enquiry and justification on the other? (MacIntyre, 1988, p. 7)

I shall be concerned with justice and rational enquiry and justification though the latter two are the more obvious concerns at this juncture. It is a mistake to think that MacIntyre is seeking a new form of understanding derived from that which was excluded by the Enlightenment. It appears to me that he is searching for what has been discarded which should not have been and which may well provide a solution to the problems that we now face.

The Enlightenment problem is a problem with two aspects, the ontological and the epistemological. What was discarded ontologically and epistemologically by the Enlightenment which was present prior to that event? Hume introduced arguments against the validity of induction and arguments separating the empirical from the normative. Both these arguments denied what had previously been accepted. Those arguments continued to resonate in the last century and to this day. Consider the argument denying the possibility of induction, an argument that goes to the core of scientific inquiry.

Mendeleev was able to predict successfully many of the properties of elements in the Periodic Table on the basis of one or two of its properties and its presence in a family by relying on the assumption that its membership of a family was the result of a common underlying structure and that, as a result, other properties could be inferred. Mendeleev did not need to know of what the common underlying structure consisted. He presupposed it and it became a possible object of inquiry at that point but there was no immediate need to know of its detail or explanation. In the same way it is possible to deal with compounds in chemistry, or with phyla in biology. Note that if the presupposition upon which Mendeleev relied is generalised to a presupposition about the nature of the physical world, it is an ontological presupposition. It is a candidate for a hard core.

The success of Mendeleev and generations of chemists and biologists are instances of inductive argument which relies upon the supposition that there are natural kinds. Do natural kinds exist and do we have methods of ascertaining them? Mendeleev, his fellow chemists and biologists without number would be thought to have extraordinary powers to make lucky guesses if such natural kinds did not exist.

The phrase 'natural kind' is commonly taken to refer to a category in nature. Butterflies, elephants, the alkali elements and igneous rocks are all natural kinds. According to Markham (1994) one of the most distinctive features of natural kinds is the richness of the structure associated with them. So, butterflies have observable properties some of which they hold in common. There are, however, as biologists delve deeper, an array of unobservable properties which butterflies have in common. The point being, that for all natural kinds, investigation tends to produce accounts of unobservable properties held in common and the development of scientific theory enables more and more sophisticated accounts to be given of the presence and structure of natural kinds. The sophistication leads to sub-kinds. And this is all by a process of induction.

The investigation of the phenomenon of natural kinds and the existence of and use of kinds constructed by society has received a great deal of attention by both philosophers and psychologists in recent years. A seminal essay by Quine (1969) appears to be the starting point of a revival of interest in a topic that had languished since the nineteenth century. Markman and Gelman (1986; 1987) have studied the possibility that human beings are either biased towards thinking in natural kinds or learn to do so very early in their lives. Quine (1969) had earlier raised the question of why human subjective categorisation of nature seems to accord so well with the actuality of nature. He noted that:

> There is some encouragement in Darwin. If people's innate spacing of qualities is a gene-linked trait, then the spacing that has made for the most successful inductions will have tended to predominate through natural selection. Creatures inveterately wrong in their inductions have a pathetic but praiseworthy tendency to die before reproducing their kind. (Quine, 1969, p. 126)

Brown (1994, p. 66) criticises this passage for suggesting that the inductive conclusions to which Quine refers are being treated as true or false. Evolutionary success would then depend upon a true inductive conclusion. But this is to interpret Quine in a particular way which suits Brown's argument. The question is whether or not the passage should be taken as saying that the inductive conclusions are adapative because they are correct or correct because they are adaptive. It appears to me that Quine is asserting that some inductive conclusions are such that they aid the survival of the species whereas others do not. They may then be talked of as right because they function successfully. They are correct because they have enabled adaptation. This is to assign right and wrong, true and false a functional value in the context of natural selection. The research of Markman and Gelman (1986, 1987) can be taken to indicate that either human beings have an innate predisposition to view the world as composed of kinds both naturally and socially constructed or they learn to do so very early on in their lives.

Kornblith (1993), following Boyd (1988, 1991) has argued that natural kinds are identified with homeostatic clusters of unobservable properties. For a natural kind to exist in a state of stability there must be a homeostatic cluster which defines and keeps in existence that natural kind. By this I take her to mean that for a natural kind to exist there must be a physical equilibrium involving a balance of functions and chemical composition. This presupposes that the world of our experience is maintained in some semblance of order so that things may continue to exist in a stable form for some observable period of time. If there are natural kinds and if their existence depends upon an ordered natural world then the investigation and recognition of the unobservables that give rise to natural kinds is an attempt to carve nature at its joints.

The alkali metals are a natural kind as are the halogens. The alkali metals combine with the halogens, common salt being a typical example. All the crystals of the compounds formed when the alkali metals combine with the halogens are indistinguishable to the observer. In observing a crystal formed by an alkali and halogen combination the observer is in the presence of a natural kind. But observable similarities are notoriously unreliable as indicative of membership of a family or a natural kind. The argument put forward by Boyd and Kornblith is that it is the unobservable structure that gives rise to the stability and the similarity that constitutes a natural kind. Mendeleev when he was working out his Periodic Table was confronted by problems arising from an incomplete list of the elements. He was aware of the observable properties of the known elements and, on this basis, could group the known elements into families. This is, however, only an indication and it is the unobservable structure which defines the kind. His mathematical key depended upon the characteristic atomic weight of each element. In the order which he knew, potassium was followed by calcium and the next atomic weight to be known was titanium. Potassium lined up with the rest of the alkali metals and calcium with beryllium and magnesium. They were clearly lined up with the right families on the basis of their observable behaviour. Titanium, next in known atomic weight, lined up with boron and aluminium to which it bore no observable familial resemblance. Mendeleev then assumed that an unknown element fitted between calcium and titanium and continued to build his table. He used the same assumption on other occasions. Mendeleev assumed an unobservable structure defining the natural kind.

His Periodic Table had some incoherence. This became even more apparent with the discovery of isotopes. Moseley, working in the context of Mendeleev's Periodic Table and Bohr's theory of the atom produced another classification.

Moseley's table was based on atomic number rather than atomic weight and differed from the pattern of classification put forward by Mendeleev in a number of instances. But Moseley was working from the same presupposition as Mendeleev, the presupposition that there was an underlying pattern, an unobservable structure defining a natural kind. Bohr's structure of the atom underpinned an argument that showed why there were gaps in Mendeleev's original table and why they had to be as they were. Bohr's theory then pointed the way to the restructuring of the table. None of this work could have been achieved in the absence of knowledge of the history of the work done on elements and in the absence of a presupposition about the existence of natural kinds and that the physical world had an underlying structure.

We have here a progression from the observable to the unobservable. A progression from observable families to unobservable homeostatic systems. We can observe the families and we can make mistakes in thinking that we have seen an instance of a natural kind. Delving into the unobservable homeostatic systems enables us to pick out the natural kinds with far more certainty. But it must also be noted that the work of these scientists depended upon their acceptance that they were researching natural kinds. It may well be that science is driven by a presupposition that natural kinds exist and that its inquiry is an inquiry into the structure of natural kinds.

The argument can now be put in the following way.

- When a population is uniform with respect to some property then an inference from a small sample or even a single case is reliable and reasonable.
- We infer from small samples and single cases constantly.
- We refuse to infer from small samples and single cases on some occasions.
- How do we discriminate between those occasions when it is reasonable to infer from a small sample or a single case and those when it is not?
- We have a natural tendency to recognise natural kinds.
- We have a natural tendency to take observable characteristics as only a rough guide to the existence of a natural kind.
- When we see evidence of a natural kind we feel able to infer from a small sample or a single case and it is reasonable to do so.
- Inductive argument is reasonable.

The above is indebted to Kornblith who states that she wishes to argue:

> When our inductive inferences are guided by our intuitive grasp of the real kinds in nature, and when we project those properties which we intuitively recognize to be essential to those kinds, our tendency to make inferences in accord with the law of small numbers serves us well. Our conceptual and inferential tendencies jointly conspire, at least roughly, to carve nature at its joints and project the features of a kind which are essential to it. This preestablished harmony between the causal structure of the world and the conceptual and inferential structure of our minds produces reliable inductive inference. (Kornblith, 1993, p. 94)

The epistemological problem posed by Hume's denial of inductive logic and inductive reasoning can now be seen to have a possible solution.

The word 'innate' has a long history. In philosophy the theory of innate ideas is normally taken to be the claim that all knowledge rests upon or consists in universal concepts of reality or objects which transcend sensory experience. Such knowledge is viewed as *a priori*. Scientific knowledge is viewed as knowledge of the physical world which is true and independently so. It is, moreover, thought to be true independent of sensory experience. There is a clear incompatibility here with empiricists who argue that sensory experience is the base from which inductions lead to scientific knowledge. The theory of innate ideas has at its base the notion that knowledge of reality is not possible directly but achieved through the medium of representations. The universal concepts and transcendent objects are looked upon as innate ideas with which humans are born. The notion of innate ideas in the theory appears to refer to one of two things. It can be taken to refer to an inborn mental image or it can be taken to refer to a disposition for the mind to form certain concepts under certain conditions.

In psychology the word 'innate' is a matter of current debate. It is usually taken to refer to the constraints on behaviour or development with which human beings enter the world. For instance, an innate predisposition to process sensory stimuli in a certain way is a constraint in the sense that it is a limitation on the range of possibilities. So, for instance, it may be argued that human beings come into the world with a built-in conception of the nature of objects, their solidity, their permanence, their extension. It may be argued that human beings come into the world with a basic conception of grammar. The point is that such conceptions are seen as constraints upon possibilities so that if it is the case that a human being is born with a conception of the solidity of objects then that human being will see the physical world in terms of solid structures and is constrained to this viewpoint. This is not to suggest that there can be no development. Development, it is suggested, can and does occur under the impact of the environment, both social and physical, and it is best to look at the situation as an ongoing interactive system between the innate constraints and the environment with the possibility that in the adult the innate predispositions may have been extensively developed.

Typically, this notion of innate is linked to an evolutionary perspective. We are, after all, animals and we tend to forget or ignore this. Darwinian evolutionary theory is the basis of modern biology and it applies to human beings as much as to butterflies. Individual living organisms are variable in their makeup. Those that are better adapted to their environment survive, those that are not do not. Thus there is positive and negative selection. It makes no sense to discard the notion of evolutionary development when thinking of the human mind. Over the course of some millions of years the human mind has developed and evolved in ways which best enable that mind to survive and develop in the environment in which it finds itself (Bronowski, 1973, ch. 13). The particular fallacy which leads to us think that there is a distinctive gap between ourselves and the rest of living things will appear again later.

Evolutionary development can be specific development for the good reason that evolutionary development results from the impact of specific environmental conditions on the living organism. In some circumstances it would be counter-productive to evolve generalised adaptive responses to the environment. But having said that there is a case for supposing that evolutionary conditions can lead to general development. Human beings, it can be argued, have developed specific adaptations such as those linked to the five senses. If, however, one argues that human beings have developed a specific adaptation enabling the grasp of grammar and a language, then as language occurs within social groupings and is generated by the environment of the social group, it could well be argued that there is a case for a generalised adaptation linked to life in a social group. The upshot of this is that there are probably specific adaptations and general adaptations in human beings. I shall follow the literature (Bjorklund, 2000) by referring to them as domain-specific and domain-general abilities which are the result of human evolution.

Consider now the modelling approach advocated by Keeves (2002). It is clear that the notion of a natural kind fits the requirements that Keeves (1997) sets for a satisfactory model. In fact it is reasonable to consider a natural kind as a model. The notion of a homeostatic cluster is also a model. Keeves (2002) considers the terms 'hypothesis' and 'model' to be interchangeable and, interestingly given the connection between the idea of homeostatic clusters and Quine, relies upon Quine and Ullian in describing the virtues that any hypothesis should have.

It is, however, in his description of the procedures for testing models or hypotheses, that Keeves (2002), following Lindsey (1995), gives an account which fits most comfortably with the notion of homeostatic clusters and natural kinds. Maintaining that there is such a comfortable fit takes the notion of modelling as developed in Keeves (2002) away from its reliance on the Piagetian undifferentiated account without necessarily discarding the distinction between the earlier Piagetian stages and formal operational thought which clearly connects with reflective thinking.

To return to the initial question. Aristotelian thought, in both its original context and as it developed through Aquinas up to the time of Galileo, approached the empirical investigation of the world of our experience through the presupposition of natural kinds. Scientists continued to practise their profession presupposing natural kinds and inductive argument while the philosophers were in denial. For instance, in 1951, Crick and Watson began to study the structure of DNA. The question was – what was the architecture? In 1953 they published their results. They argued the structure of DNA to be a right-handed helix in which each link is of the same size and turns through the same rate between links. DNA was the fundamental biological invariant but the composition of DNA varied from class to class. Their work presupposed natural kinds and induction.

It is arguable that, of any theory, it is very probably approximately true if it is the best explanation of empirical observations provided that it, the theory, was not produced in the first place in order to explain those observations. A theory produced to explain certain observations is, necessarily, going to have some success, but that success is irrelevant to the truth of the theory. On the other hand a theory which has not been concocted to explain some observations may, by chance, do so successfully. The more observations a theory can explain the more we may be inclined to accept it as warranted. Where the observations appear drawn from different classes of things and are explained then this strengthens our belief in the theory. In essence the argument is that the theory is warranted because of its performance. Explaining a range of observations which appear drawn from different classes of things creates the assumption that the range encompasses a class. Once again we return to the presupposition of a common underlying structure.

This puts us in the situation where practising scientists presuppose a common underlying structure and use inductive reasoning while some philosophers of science deny that it is possible.

The argument that we see the world in natural kinds clearly presupposes that there exists a world independent of ourselves and that in seeing and developing our understanding of the natural kinds that exist in the world we are carving nature at its joints. It would appear that the argument presupposes ontological reality and our ability to know it and to reflect on it.

The argument has been developed through examples drawn from the natural sciences but there is no reason in principle to suppose that the argument does not apply in the social and behavioural sciences. In what sense does the psychologist practising psychology see the world in natural kinds? The presumption must be that he or she does.

The explanation of what the human being is doing or has done will be in terms of an account of behaviour not of action for action indicates voluntariness. An explanation in these terms rules out the possibility of reflection being a significant part of what it is to be human.

In turning to put a positive case the task must be to argue for a particular model of what it is to be human and to place the concept of reflection squarely within that model.

CHAPTER 9

SCIENCE AS PRACTICE

It is from MacIntyre's concept of practice-narrative-tradition that I start a positive case.

A practice is:

> … any coherent and complex form of socially established cooperative human activity through which goods internal to that form of activity are realized in the course of trying to achieve those standards of excellence which are appropriate to, and partially definitive of, that form of activity, with the result that human powers to achieve excellence, and human conceptions of the ends and goods involved, are systematically extended. (MacIntyre, 1981, p. 187)

Some important points can be made about this definition. A practice is a social, cooperative human activity. It is not something in which an individual can participate alone. Participation necessarily involves social exchanges and cooperative activity. That is, practice involves not people doing things to others but doing things with and for others. What normally counts as a practice can be reduced to an individual pastime by the individual or the group or the culture who see it as a solitary activity. Teaching is a practice when it is seen as a socially cooperative activity in which the teacher does things with and for others, not just other teachers but the students. It starts to cease to be a practice when the teacher sees himself or herself, and the school and the authorities see the teacher, as performing a solitary task in the classroom. Yet this has been and still is a common viewpoint. Teachers and scientists may be engaged in individual actions but, if teaching is to be seen as and be a practice, the actions must be seen and understood as connected with and dependent on others who are engaged in the same practice and that practice is to be seen within a tradition and a narrative. The actions have to be seen as a form of activity which has developed over time into a complex activity which has become socially established.

This is not to reduce a practice to group activity. A group activity can be made up on the spot by any so minded group. It starts to take on the features of a practice when the group starts to define the activity and produce criteria by which achievement in that activity is measured. A ball game is a simple case in point. It may start by kicking a ball about. When it becomes competitive and rules start to be produced for the proper playing of a game, and skill starts to be recognised, then it becomes a practice as it becomes a complex, socially cooperative form of human activity. Seeing the kicking of a ball develop into football one is seeing the history of the creation of a practice. Each step along the way is a response to and examination of the current state. Tactics develop over the years and the current players need to have some awareness of the former tactics if they are to attempt excellence. What is true of a simple example like football is true of complex activities such as the practice of science or of teaching. Each has to be seen in the context of its history for the present state is a result of responses to particular events and ideas in that history. This is to say that a practice is under constant scrutiny and modification. If this does not happen the practice ossifies and dies.

A practice then has an end goal of the constant improvement of the practice. Where the practice is the playing of a game such as chess then the improvement is sought in the extension of knowledge about gambits and such. Where the practice is the playing of a game like rugby improvement may be sought by modification of the rules. The modification cannot be such as to change the character of the game otherwise you end up with a different game – Gaelic football and Australian rules are examples, as indeed are the many codes that go under the name of football throughout the world.

Where the practice is a form of inquiry such as science then the constant improvement may be sought in the products of the inquiry and the mode of inquiry. Where the practice is an activity such as teaching then the constant improvement may be sought in the mode of the activity and the results of the activity. The constant evaluation of the practice is necessarily connected to a continued attempt to achieve that excellence through which the goods internal to the practice can be realised. There are two types of goods involved here (MacIntyre, 1981). The first is the good which flows from the excellence of the products and the excellence of the performance. There is, however, a further and perhaps more important good which flows from the practice. This is the good that flows to the practitioner in participation in the practice. A scientist practising science excellently, or a teacher practising teaching excellently, necessarily experiences that good which flows from participation in a certain sort of life. A student learning science starts to experience that same good. These goods are the effects on the scientist or the teacher or the student as a person. Participation, in the form of life which is the practice, affects the mind, the imagination, the values and the humanity of the scientist or the teacher or the student. The effect is open only to those who participate, with the greatest effect open to those who participate well. Note that the effect is not open to those who only learn about science for they are onlookers rather than participants. MacIntyre and others have argued that it follows that only those who have participated are in a position to appreciate and recognise the values internal to the practice. I am dubious and will return to this issue later. The effect of participation is such that it may properly be seen to be connected to the achievement of human flourishing. However, to claim that is to assume that there are no evil practices, an issue to which I shall return while continuing with the assumption.

Because a practice is connected to human flourishing it has within it connections to values. However, particularly in the case of those practices which are inquiries or partly inquiries, there is a connection with knowledge. The scientist is seeking knowledge, the teacher is seeking knowledge as well as attempting to pass it on. Outside of the practices it may be possible to have some simplistic or technological understanding of the knowledge produced by the practice but it is only within the practice that the knowledge can be seen and understood in its social and historical context and then not necessarily by all practitioners where practices have a number of specialities. There is then a connection between knowledge and values, between knowledge and human flourishing that will be explored later but it would seem to involve the generation of new ways of perceiving the world of our experience through initiation into and participation in a practice. So a practice may be intrinsic to or constitutive of a life. Teachers attempt to initiate students into practices that are constitutive of a life. It follows that teachers or curriculum designers should not attempt to limit initiation into practices such as science to those who would follow them as professionals. The tentative mentioning of values and knowledge by the adherents of the strong program and Longino and Allchin does not go far enough. They are hindered by working within the tradition of the Enlightenment

The above has been written as though practices can be identified discretely. While they may be, it must be remembered that the situation is far more complex. Science as a practice is composed of many sub-practices – biology, chemistry, physics and like disciplines. Chemistry is composed of many sub-practices – inorganic, organic, physical and like fields. Each sub-practice of science may call on a wide variety of practices and sub-practices from within the community in which the science is practised. Within physics there are a number of sub-practices all of which contribute to the maintenance of a tradition and the same is true of other practices.

In everyday discourse when the word 'practice' is used we tend to think of somebody doing something. A tennis player may practise by hitting balls against a wall. We think of teaching practice as a student teacher exercising what they have learnt about teaching a group of students. This use of 'practice' focuses on the individual and his or her performance. When we use the phrase 'X practice' where X may be legal or architectural or dental we are using the word 'practice' in the same sense. We tend to think of it as a person performing, the usage is always about the individual.

It is also a usage which tends to focus on the here and now. It is about a tennis player hitting balls and refers to a specific moment in time; the student teacher at a specific moment in time or the lawyer, the scientist, the architect. I do not mean here to give the impression that the length of time is a second or a minute but rather that the time is a time with no history. We tend not to think that we may be referring to a tennis player in the context of his career or the student teacher in the context of theirs unless we specifically qualify the usage. With the lawyer we are talking about a person performing a series of actions that can be identified as related to certain activities in society but, unless we deliberately chose to, we do not think we are referring to something in its historical context whether that history be of the individual or society.

It is also a usage which focuses on the technical. The tennis player is perhaps trying to improve his or her striking of the ball. The student teacher may be practising the management of a group of students and the imparting of knowledge. The lawyer is going about his or her business as is the architect. The emphasis is on technical employment of skills which are either being learnt or have been learnt.

We pick out the same three points when we talk of a 'practising scientist' or a 'practising doctor'. Our usage is individualistic, technical and ahistorical (Dykstra, 1997). It is an image of the professional exponent of a set of skills into which only a few have been initiated.

This usage of practice is at odds with the notion of practice which I have put forward above. This raises the question of whether or not this results in problems for those wishing to practice in MacIntyre's sense.

A number of points may be made, some of which will require later development.

- An individualistic and ahistorical notion of practice is incompatible with the notion of practice as a complex, social cooperative human activity.
- If practices are seen as individualistic, ahistorical and technical then initiates into a practice will be taught as though that is what it is. In a school a teacher or administrator or curriculum designer who sees practice as individualistic, ahistorical and technical will not see the subjects that are taught for what they are.
- Practice seen as individualistic, ahistorical and technical will hinder the achievement of those goods internal to practice.
- Practice seen as individualistic, ahistorical and technical will hinder the achievement of human flourishing.
- Practice seen as individualistic, ahistorical and technical will, in teaching, be seen as a performance in which one person does things to others rather than with others.

MacIntyre (1981, p. 200) allows that there may be evil practices but asserts that he is not convinced that they exist. To my mind, there are three points that can be made in this connection.

- If virtues such as justice, prudence, temperance and fortitude are necessary for the practitioner then the nature of those virtues and their possession and use necessarily provide a self-correcting influence on the practice. This is not to say that a practice cannot be evil but it is to say that it is not a practice as defined for it cannot be an expression of the virtues. The argument that the virtues are a requirement of a practice will be developed later.
- A practice, as defined, cannot be ahistorical, individualistic and technical. The individualistic tenor of Western society tends to have the effect that practices are seen as individualistic performances. In addition there is also an element of seeing practices as ahistorical and technical. Evil 'practices' exhibit some or all of those characteristics. For instance, the expert confidence trickster may be thought to be an evil practitioner. He or she is, in their 'practice', anti-social, that is, extremely individualistic.
- The notion of the *telos* of human flourishing is the criterion to which all putative practices submit and this is incompatible with evil practices.

I shall take the view that practices, as defined, are inherently good.

The practice of science is a 'coherent, complex form of socially established cooperative human activity'. Science is a human activity. It cannot be ahistorical or asocial. To be initiated into the practice that is science is to be initiated into a practice that has a history and that initiation cannot take place except in the understanding of that history. All scientific work is a response to a problem which has a history both intellectual and social and any attempted solution becomes a part of the history. In this way all scientific work is adding to and, at times, reconstructing its history.

> It was important when I characterized the concept of a practice to notice that practices always have histories and that at any given moment what a practice is depends on a mode of understanding it which has been transmitted often through many generations. (MacIntyre, 1981, p. 221)

Chemistry and alchemy went hand in hand in for a long time. If a point of parting is to be chosen then the publication of Boyle's 'The Sceptical Chemyst' in 1661 is commonly chosen, but for some time after this those with a chemical bent continued to seek the secret to transforming ordinary materials into gold or other valuable material. The domination by Sweden of the matchmaking industry has its origins in the search to turn urine into gold. Phosphorus not gold was the result. Throughout the seventeenth century and up to the end of the eighteenth century what was to become the new subject of chemistry was hardly a coherent, socially cooperative human activity. Priestley is commonly credited with the discovery of oxygen in 1774 and Davy with the discovery of chlorine in the early years of the nineteenth century. In fact Scheele discovered both but died in 1786 without being able to get his results published. To understand chemistry as it is now it is necessary to grasp sufficient of its history. The same is true if the initiate is to come to an understanding of chlorine or oxygen, their history must be a part of the understanding. For the practice of chemistry and the elements chlorine and oxygen are all answers to problems which arose and have to be understood in that context. Certainly it is possible to go ahead without any knowledge of the history of chemistry or oxygen or chlorine, but then what exists is a technician not a practitioner, and even then it is a limited technician.

CHAPTER 10

SCIENCE AS NARRATIVE

Consider a stranger drinking a liquid from a container. We can describe the mechanics of the situation. If we have sufficient physiology we can describe what is happening beneath the exterior of the body that we see before us. We can give the circumstances a cultural context by recognising that the container is a wine glass and that the liquid is red. But while we can describe the mechanics, and by looking at the cultural context we can guess that the stranger is drinking red wine, we cannot with any surety give a description of the human action which is taking place. The act of drinking the red wine may be described as:

> … an act of self-indulgence, an expression of politeness, a proof of alcoholism, a manifestation of loyalty, a gesture of despair, an attempt at suicide, the performance of a social rite, a religious communication, an attempt to summon up one's courage, an attempt to seduce or corrupt another person, the sealing of a bargain, a display of professional expertise … (Ayer, 1964, p. 7)

The point being that the stranger is the final authority on what he or she is doing. How he or she sees his or her behaviour is what makes the action the action that it is. The mechanics are the same, the action differs. In the same way measurement of human behaviour may capture the mechanics of the circumstance but misses the humanity of the behaviour. But if we are told by the stranger that it is one of the possibilities mentioned by Ayer then we need to know more if we are to understand it fully. We need to know how it appears at this moment in the life of that person in this society. The society in which a person lives serves as the setting for his or her life. It consists of institutions, norms, practices, language and all the myriad items that make and eventually distinguish one society from another. It is against this background that we are able to understand the actions of a person. Socrates deliberately drinking a poisoned cup is understood in a different setting and therefore differently from someone quaffing wine at a barbecue. The actions may look the same or very similar but the settings and the intentions of the drinkers are different and the actions cannot be understood for what they are apart from the setting and intentions of the actor. This is why on occasion we ask – what do you think you are doing? or, of a fellow spectator, what is he thinking of? The point being that no understanding of human action can be ahistorical or asocial. Moreover for the individual no response to or consideration of the question – what am I to do? – can be ahistorical or asocial. All our actions and decisions are necessarily seen by us in the context of our existence, in the context of what MacIntyre calls the narrative of our lives. The death of Socrates is to be understood against the narrative of his life and his friends had to be reminded of this. The question of what am I to do is a question answered by considering who I am. If that is true of us then it is also true of those who would attempt to understand what we do and who we are.

What is true of the attempt to understand the individual is also true of the actions of societies and nations. The English viewing the events of 1789 in France could see the events and, in a sense, describe them. They did not necessarily understand them nor the causes that led to them. If there is to be an understanding of the events of the French Revolution then the events of 1789 must be placed in the context of an historical narrative. The historical narrative that might have been written in 1815 could well be different to that written in 1900 and that written in 2000. The narrative is written with the information and ideas at the disposal of the writer at the time. Historians recognise that accounts may change as information changes and ideas and analyses change. The point is to write the best narrative using the information and knowledge at the time of writing. This information and knowledge will include what is known of Marie Antoinette, Robespierre, Louis and the culture and society at the time. Such information and knowledge can well change over time. The historian also brings to the writing of the narrative various analyses or models which operate somewhat as templates.

Science is a human activity and, in order to understand the work of scientists and what they produce, it must be seen in its historical context. The narrative of the science must be written. The revolutions in scientific thought are the places in which the narrative is rewritten. The rewritten narrative enables the reader to see where the past was awry and where the future might lie (MacIntyre, 1977).

For a society to exist and continue to exist it must be possible for the members of that society to give an intelligible account of their own actions and those of others on sufficient occasions. Without accountability society descends into confusion for everything becomes unpredictable and unintelligible. For society to cohere sufficiently, for its existence to continue without stagnation, it must be possible for the members of that society to predict with sufficient confidence a large range of actions of the members of that society. To the extent that any society coexists with others then the same dictum applies to a grasp of the actions of another society. Societies like individuals do not exist in social vacuums. An intelligible account depends upon a grasp of the narrative. This is true of human life and is true of human practices such as science for the concept of narrative applies both to the inquiry and the inquirer. A failure to grasp the narrative of another society can bedevil foreign policy.

As human beings we learn to understand the actions and predict the behaviour of others. Predictions can be wide of the mark. One of the major contributions of social science since the mid-nineteenth century has been the improvement in our understanding of human behaviour and action. We now have a better and improving grasp of the underlying probabilities. If there is a problem, it is the failure at times to see individuals, groups and societies in the context of their narrative and an inclination to treat them as individual objects to be examined in an historical vacuum.

The concept of the narrative is in contrast to the tendency to explain human action by attempting to categorise individual actions as belonging to certain roles. In the first place, it is not intelligible to talk of individual actions and, in the second place, it is not intelligible to talk of distinct and separate roles. It is true that it is possible to see human beings as compartmentalising their lives so that they appear to wear different so-called hats on different occasions, but I would argue that this is a function of a Western tendency to see human beings as individuals and their acts, as far as possible, as isolated instances. It is a model with both positive and negative analogy. In the same way the history of science may be seen as being the story of isolated individuals, usually men, performing isolated experiments. It is at this point that the work of the sociologists and social psychologists is a useful corrective. Of even more use might be a consideration of the approaches of other cultures. I have in mind the Confucian culture of East Asia.

CHAPTER 11

SCIENCE AS TRADITION

Matthews (1994) in a chapter 'History of Science in the Curriculum' sets out some of the reasons that have been given for including the history of science in the science curriculum.

> (1) History promotes the better comprehension of scientific concepts and methods.
>
> (2) Historical approaches connect the development of individual thinking with the development of scientific ideas.
>
> (3) History of science is intrinsically worthwhile. Important episodes in the history of science and culture - the Scientific Revolution, Darwinism, the discovery of penicillin and so on - should be familiar to all students.
>
> (4) History is necessary to understand the nature of science.
>
> (5) History counteracts the scientism and dogmatism that are commonly found in science texts and classes.
>
> (6) History, by examining the life and times of individual scientists, humanizes the subject matter of science, making it less abstract and more engaging for students.
>
> (7) History allows connections to be made within topics and disciplines of science, as well as with other academic disciplines; history displays the integrative and interdependent nature of human achievements. (Matthews, 1994, p. 50)

Matthews argues that underlying (1) is a belief that the better comprehension of science is necessarily historical. However, his arguments do not support the notion that it is necessarily so. That is, that the narrative of science is a necessary part of practising science. Boltzmann was not in a position to solve problems with classical mechanics without understanding the problem. The problem is defined by its narrative, its history. Matthews' arguments, with one notable exception, are confined to attempting to demonstrate that the best way of acquiring an understanding of the nature of science and scientific concepts is through the addition of an historical element in the science curriculum.

The exception is a quotation from Fleck (Matthews, 1994, p. 50) which asserts that there is no understanding which is ahistorical. This is an assertion about the nature of understanding and it goes far beyond the claims that we achieve better understanding when an idea is placed in its historical context.

One approach, exemplified by (7) above, that has been very influential was that put forward by the Harvard Committee Report (1945). The approach put forward is summarised in:

> Science instruction in general education should be characterized mainly by broad integrative elements - the comparison of scientific with other modes of thought, the comparison and contrast of the individual sciences with one another, the relations of science with its own past and with general human history, and of science with problems of human society. (Conant, 1945, pp. 155-156)

Conant saw science as an abstract study that was dependent upon and partially constituted by tradition. This, he thought, tended to make science a difficult subject to study. Part of the solution lay in the broad integration of science with other areas and modes of thought and, in particular, with history. The history was to be not just the history of science but science in the context of human history and the society in which it was practised.

This is an attempt to solve some of the problems of learning science. History is put forward as something which illuminates and integrates. It is not seen as something which is a necessary component of the practice of science and initiation into that practice. In saying that tradition makes the learning of science difficult Conant appears to be viewing a tradition, such as classical mechanics, as a tradition that stands outside its own history. But this cannot be. A tradition does not come into being overnight but grows and is extended as it is used, criticised and modified. Classical mechanics did not leap in full bloom from Newton's mind but developed and changed over decades until its serious flaws became apparent. In consequence, if we are to understand a tradition it must be viewed in the context of its history. But it is more than that. The current point in any tradition is the point that has been reached by use, criticism and change. Therefore the current point can only be understood and evaluated against the past to which it was a reaction. The tradition is the justification of the current state.

Returning to the list provided by Matthews, if we discard (2) with its dubious connection to a Piagetian model of development then all the reasons are reasons which are external to the practice of science, that is, they point in some way or other to an external benefit. It could be argued that this is bound to be the case given that science and history are treated as separate entities and the problem is to combine them usefully. Some will attempt a combination of social history and science, others political history, others scientific ideas. There appear to be two views, those who adopt an internalist approach and those who adopt an externalist approach (Matthews, 1994). My view is that all these approaches are external in an important sense. And, worthy as they may be, they are still external approaches and because of that miss a crucial feature of science and traditions.

> A living tradition then is an historically extended, socially embodied argument, and an argument precisely in part about the goods which constitute that tradition. Within a tradition the pursuit of goods extends through

> generations, sometimes through many generations. Hence the individual's search for his or her good is generally and characteristically conducted within a context defined by those traditions of which the individual's life is a part, and this is true both of those goods which are internal to practices and of the goods of a single life. (MacIntyre, 1981, p. 222)

This is an account of a tradition as a search which takes the form of an inquiry and an argument. Science as a tradition is science as an inquiry into the nature of the world in which we live. What counts as progress in that inquiry is always a matter of debate and argument not simply about the results but about the methods and the nature of the inquiry. What counts as the good in that inquiry is not only knowledge and warranted belief but also the contribution that the inquiry makes to human flourishing both externally, and, internally to those participating in the inquiry.

MacIntyre expands upon this definition in a later work:

> A tradition is an argument extended through time in which certain fundamental agreements are defined and redefined in terms of two kinds of conflict: those with critics and enemies external to the tradition who reject all or at least key parts of those fundamental agreements, and those internal, interpretative debates through which the meaning and rationale of the fundamental agreements come to be expressed and by whose progress a tradition is constituted. (MacIntyre, 1988, p. 12)

There are clear links with the ideas of Lakatos (1978a). The fundamental agreements may change with time. In physics there has been change from Aristotle to Galileo, to Newton, to Einstein. These fundamental agreements (in Kuhnian terms, paradigms) changed under the pressure of internal debate and external attack. However, it is a mistake to think that such change is always progress for sometimes there are side tracks that lead nowhere and ideas that are sterile.

An important concept within MacIntyre's approach is implicit in his account of change in tradition. If the change is brought about by a combination of external criticism and internal debate then the inference is that there are internal criteria by which the internal debate progresses and that external critics in some manner are able to influence internal agreements. If this inference is considered alongside the notion that there are no independent criteria which stand outside all traditions and to which disagreements are referred, then, there must be some procedure by which external criticism has purchase. This is a necessary result of the denial of independent criteria and a procedure has been sketched and will be developed further later.

CHAPTER 12

SCIENCE AS FUNCTION

The concept of 'science' is a functional concept. That is, we define 'science' by way of the purpose or function which it is expected to serve. This however needs some qualification. While physics and biology are considered to be sciences they have distinctly different purposes or functions in relation to their inquiry into the physical world. It would therefore be better to remark that the concepts which are grouped under the generic title, sciences, are functional concepts which are defined by way of the purpose or function which they are expected serve. Another qualification is needed. It would be better to say that the sciences are functional concepts which are defined by way of the purpose or function which they are characteristically expected to serve for it is clear that there is some disagreement at times over purpose or function.

Functional concepts necessarily bring into play the notion of what is good and what is bad in terms of the service of the function. There is always the question of whether or not the function or purpose is being served well or ill. A piece of metal may function as a knife and then there is the question of whether or not it is a good knife or a bad knife. The initial consideration in calling a piece of metal a knife is to decide whether or not it is sensible to apply that functional term to the piece of metal. If the term 'knife' is applied to a broad, blunt, soft piece of metal then an observer can quite sensibly object to the use of the term. The observer may well point out that the object could not function as a knife for the function of a knife is to cut and that piece of metal would not cut soft butter though it might spread it. But this should not be taken to mean that it is only after the initial criteria for the appellation 'knife' are satisfied that there is consideration of whether or not it is a good knife or a bad knife. The criteria by which you determine whether or not a piece of metal is a good knife, a bad knife or no knife at all are the same. An object is a knife in virtue of its ability to cut. The use of 'virtue' links the definition to the notion of excellence. A knife cannot be defined independently of the idea of a good knife.

If we wish to understand the eye we need to know how it functions such as how it acts in relation to light. This is the function of the eye at one level, the level of how it is working. The function of the eye in the body has to be understood in relation to the welfare of that body. Without the contribution of the eye the body would be unable to function properly, it would be crippled. The eye cannot be defined independently of the idea of it functioning properly and this incorporates the notion of its characteristic function.

Miller is at odds with this:

> ... in the case of a productive activity like architecture or farming, or in the case of an intellectual activity like physics, there is an external purpose which gives the practice its point and in terms of which it may be judged – in the cases in question we might say, respectively, the creation of attractive and comfortable buildings, the production of food for the community, and the discovery of scientific truth. (Miller, 1994, p. 250)

Miller then argues that the good of being an excellent physicist is not constituted internally to the practice. In my view he is mistaken. In asking the question whether or not a practice is to count as physics you necessarily weigh performance against criteria and ask whether or not the performance measures up to the criteria. So, physics, *qua* physics, cannot be defined independently of the concept of good physics and bad physics. It follows that the criterion of something being good physics and the criterion of something being physics are not independent of each other. And it follows that the criteria by which the performance of a physicist are judged are not independent of the concepts of good physics and bad physics. Miller's mistake may lie in the view he takes of the discovery of scientific truth as being a purpose external to the practice of physics. I would argue that it is partially constitutive of what is meant by physics.

O'Neill makes a related point to Miller quoting Aristotle:

> There are a number of arts in which the creative artist is not the only, or even the best, judge. These are the arts whose products can be understood and judged even by those who do not possess any skill in the art. A house, for instance, is something which can be understood by others besides the builder; indeed, the user of a house – or in other words the householder – will judge it better than he does. In the same way a pilot will judge a rudder better than a shipwright does; and the diner – not the cook – will be the best judge of a feast. (Aristotle, 1948, 1282a, 17ff., p. 147)

O'Neill follows this by saying:

> MacIntyre's account of judgement fails to acknowledge this point. Thus he writes of 'internal goods' that they are internal 'because they can only be identified by the experience of participating in the practice in question. Those who lack the relevant experience are incompetent thereby as judges of internal goods.' This is false. As Aristotle notes, there are many practices in which it is not the participant but the user who is the best judge of its products...Indeed internal practitioners often have their judgement distorted by their appreciation of the qualities of performance as against the product. (O'Neill, 1993, pp. 141-142)

I am at odds with both O'Neill and Aristotle.

A cook engaged in practice seeks to balance the dish and the assortment of dishes. Each cuisine of standing has a different way of doing this. For instance, Chinese cuisine seeks to balance colour, aroma, flavour, shape and texture according to their yin/yang qualities and cook the ingredients in a very short time so that they retain their natural flavours, texture and nutritional value. In these respects it is similar to that other great cuisine, Italian. The cook, if he or she is good, gets it right. Only another initiate into the practice can appreciate the skill and thought that has gone into producing the dish or the meal. The diner appreciates the result depending upon his or her preferences and tastes. There will be diners who cannot appreciate the flavours and the balance, there will be others who can. However, they all appreciate them, in so far as they are not initiates into the practice of cooking, in terms of their own personal preferences and tastes. These are two different forms of judgement for they relate to two different sets of criteria. One set of criteria are internal to the practice, the other internal to the diner but external to the practice.

The shipwright builds a dinghy rudder. A rudder has a certain function. It must steer the dinghy effectively, produce the minimum amount of drag and the maximum amount of lift, be as light as possible commensurate with the strength required. A sailor using the rudder on a dinghy feels the rudder in the water and how it works for him or her. The personal preference and feel of the sailor determines how the rudder is judged. The sailor may prefer a heavier feel or a less or more sensitive rudder. The merits of the rudder for the sailor lie in personal criteria that can best be expressed by saying that the rudder feels right for him or her. Once again there are criteria internal to the practice and criteria internal to the user but external to the practice. And this argument can be made in relation to all examples that are derived from O'Neill's use of Aristotle's argument. But it must be added that those practices which produce something to be used by someone who may be outside the practice must seek and consider the judgement of the user. This may on occasion lead to a modification of the practice.

The set of criteria are clearly factual. In using the criteria to assess whether or not this is good physics the move is from a factual premise to an evaluative conclusion, going from 'is' to 'ought', a move which, from the time of Hume, has been thought to be logically invalid. It is not, and such a move is made all the time with functional concepts. The question for science is twofold – what function does the practice of science serve which is internal to that practice, and what function does science serve which is external to the practice of science. The first may be answered by considering the human activity which is science and the second will, necessarily, be affected by the culture within which the science is practised.

However, if the activity of science is thought of only in terms of the pursuit of knowledge in the context of a distinction between fact and value, then this will carry with it the notion that science is value-free or value-neutral where values are thought of as external to the practice of science. This denies the contribution of a practice such as science to human flourishing and fails to realise that this contribution is internal to the practice of science.

It is arguable that practising scientists, particularly those practising in the field of biology, have not discarded the teleological approach which derives from Aristotle and was alive prior to Galileo. Yet the philosophers of science did. Where people who have been influenced by Hume construct the means by which people are taught science then there is the distinct possibility that what is taught is not science. This is a danger to which I shall return, but there is more to be said about the effect of Hume and the supposed dichotomy between 'is' and 'ought'.

A major tenet of the majority of post-Galilean thinkers was that no evaluative conclusion can be drawn from factual premises. This view was held by Hume, Kant, Mill and their successors to the present day. In the sphere of ethics this has the effect of divorcing moral argument from factual propositions. In the sphere of science it has the effect of divorcing evaluative propositions from empirical propositions. It leads to the position where it is not possible to draw from factual premises the conclusion that this is 'good science', for that is an evaluative conclusion. This leads to an attempt to produce norms for science which are distinct from other norms and internal to the practice of science. In this way it is hoped that what counts as good science can be determined without having to go beyond the criteria of the practice which is science.

It is clear that scientists make judgements about what counts as scientific progress. Mendeleev was praised for his work, as were Thomson and countless others. So, in one sense, for the community of practising scientists the notion of what counts as good science was not lost. But the notion of good science is not a simple notion. Certainly the recent history of science shows scientists making further and further advances in their understanding of the physical world; however, there is, and has been since Galileo, constant debate about the ends which science serves.

> Scientific knowledge is proven knowledge. Scientific theories are derived in some rigorous way from the facts of experience acquired by observation and experiment. (Chalmers, 1976, p. 1)

> The different branches of scientific enquiry may be divided into two major groups: the empirical and the nonempirical sciences. The former seek to explore, to describe, to explain, and to predict the occurrences in the world we live in. (Hempel, 1966, p. 1)

> Scientific knowledge underpins modern technologies which enable all of us to survive and which can enhance or threaten the quality of our lives. A general understanding and appreciation of science is essential for all citizens. (Matthews, 1995, p. 19)

> … science is pragmatically useful, gives us some understanding of important physical and metaphysical aspects of the world, and to some degree follows a critical ideal which it is important to understand and whose spirit is worth emulating. (Couvalis, 1997, p. 8)

These words are illustrative of the debate about the ends of science. The study of science is about knowledge, its acquisition and its use. It is about rigorous inquiry, about technology, about usefulness. It is about an ideal way of inquiring. It is about some or all of these and more. Science is useful in that it enables human beings to manipulate the world of their existence. That manipulation may be towards ends which are assessed as valuable or non-valuable. Or it may be the case that the question of ends does not arise, for the ability to manipulate is an end in itself when what is valuable is matter of personal preference.

I shall argue that all of these assertions miss a crucial point about the internal goods of science, or indeed any practice.

CHAPTER 13

SCIENCE AND CULTURE

It is not uncommon in Australia, the United States and other Western countries to find something akin to the following in science curricula statements:

> Every culture has its own ways of thinking and its own world views to inform its science. Western science is the most dominant form of science but it is only one form among the sciences of the world. (South Australian Department of Education, 2003)

The view put forward appears to be that different cultures can have different sciences and that these may or may not be commensurable. In the extreme case other cultures may have different sciences and the truth value of these different sciences is internal to the culture and the cultures and their sciences are incommensurable. Ontological and epistemological problems lie at the base of this view.

I accept, following MacIntyre, that science is a practice and that a practice is 'any coherent and complex form of socially established cooperative human activity ...' (MacIntyre, 1981, p. 187). Longino (1990), moving on from Barnes, Bloor and Latour, has emphasised the social activity which is science and attempted to describe firstly, the values which are inherent in that practice and, secondly, the values which society produces and which affect the practice. Longino set out the characteristics which, in her opinion, the practice of science must have.

There is no doubt that the culture in which scientists practise science affects their practice. But this can be conceived in two ways. On the one hand there is the view that people belong to a community and cooperate for the purposes of mutual advantage. This is the view of the individual within a community, within a culture, and gives primacy to the individual. On the other hand there is the view that cultural, social and community circumstances determine in some measure the self, and that the identities of individuals are constituted by the culture and the community of which they are a part. This view gives primacy to the community. Consider the two views in turn.

THE INDIVIDUALIST VIEW

The individualist view is one that is linked to the Enlightenment. The Enlightenment formed a tradition, which had at its core, certain views of the relationship between fact and value and certain views of the individual. The Enlightenment brought with it an emphasis on the individual and individual autonomy and choice. Watt (1989) makes useful distinctions between altruistic individualism, egoistic individualism, ontological individualism and normative individualism as they appeared in the last decades of the twentieth century and briefly traces their historical origins.

Altruistic individualism rests on the Kantian premise that all individuals are ends in themselves and are to be valued as such and be free to develop their uniqueness without unnecessary interference. Egoistic individualism rests on the premise that human beings are unilaterally in pursuit of their own interests in competition with other human beings. Ontological individualism rests on the premise that human beings are the fundamental identities from which societies, cultures and groups are derived. Normative individualism rests on the premise that individual human beings are the fundamental source of value. Watt points out that these distinctions are not mutually exclusive in that any particular notion of individuality could well combine some or all of these distinctions. Watt gives instances of the variety of current standpoints:

> • a view of the relationship between individual people and society, in which individuals have prior reality and ultimate value, while social institutions have a secondary, derivative status, and exist only for the benefits they provide for individuals;
>
> • a preference for a social contract model of society, representing it as an arrangement devised by pre-existing independent individuals, rather than a model of society as a living organism incorporating individuals as its compound cells or organs;
>
> • the conviction that individual uniqueness, initiative and autonomy are more important than group identity, conformity and solidarity; and
>
> • the tendency to perceive the most basic relationship among people as one of competitive self-interest rather than of solidarity and mutual support. (Watt, 1989, pp. 1-2)

It is evident that a powerful expression of Western individualism in its current phase is that of an extreme of individual competition and self-interest. This, as Watt points out, harks back to the ideas of Adam Smith and Spencer. This cannot be taken as the sole meaning of individualism at present. However it is possible to say that at the core of individualism lies the notion that the individual is the ultimate worthwhile entity and that an individual must, to retain this individuality, be recognised as having the right to do, without interference, only that which they have decided to do and believe only that which they have independently judged to be the case.

Individualism is a complex family of views in which any one or a combination may, at a particular time, dominate. At present it would seem that there is an emphasis on the self, personal achievement and personal wants and desires. In such a tradition there will be a tendency to see the interests of the individual as taking priority over those of the community in which that individual exists. This might be put as a presumption that the individual's interests and rights prevail unless and until it can be shown that they should be curtailed in the interests of the community. The presumption is that you are free to do as you wish until and unless it curtails the freedom of others and then there must be an assessment of the circumstances. There is a presumption that in matters of value, and hence human goals, it is a matter of individual preference and choice. Communities, including that of the state, are to be arranged and administered on the presumption of minimum interference with the individual.

> In its Lockean formulation, the individual stands at the centre of liberalism, alone, as the best rational judge of one's self-interests and must be permitted to act freely to achieve them. Like-minded individuals can freely associate themselves as interest groups to pursue shared corporate interests. The only moral injunction against the pursuit of self-interests is that it should not be at the expense of others. The social is thus conceived in the negative. (Chua, 1995, pp. 185-186)

This world is a world in which the rights of the individual are fundamental and may be pursued or not as the individual wishes. The pursuit of rights may be aggressive and confrontational. This is accepted for the pursuit is fundamental to society. Civil and political rights take precedence over social and economic rights. Morality is a matter of individual preference, a matter of individual wants and desires. That individuals will have different wants and desires is expected and there are thus the seeds of dissension among people. Manipulation is the means by which people are persuaded to do that which you wish them to do.

Social institutions tend to be looked upon as merely means to enabling the individual to achieve his or her particular goals, they have no value other than an instrumental value. The political and social idea of democracy is conceived in terms of freeing individuals to achieve their goals. The identity of the self is thus construed in terms of personal achievement. How then does the individual measure this personal achievement? One measure is against the achievement of personal goals – to what extent have they been realised? Another measure is against other individuals. An individualistic society is one that tends be to a competitive and acquisitive society. Individuals measure their worth and identity against others in terms of their success in competing with others and in terms of the acquisition of that which is thought worthwhile. In the situation of a market economy this will be the acquisition of material goods.

Just as people living within that tradition will, in order to function within that tradition, live their lives in a certain way so a scientist practising within the tradition of the Enlightenment will tend to be affected by that tradition. The tendency for scientists working in the tradition of Western science will be for them to see themselves as individuals seeking individual goals and achievements though they may join with others for mutual advantage. As members of society they will value and pursue their civil and political rights and, where those rights are important to their practice, they will be pursued in that context also. This may be at the expense of others. Individual achievements are seen as the possession of the individual gained in competition with others. Intellectual property and patent law will be of concern to them. The view that knowledge is a fund for all practitioners is curtailed. The history of science in such a tradition is told in terms of individual genius and endeavour. All that has been said above is an expression of the constraints that impinge on the individual scientist in virtue of his or her work within the Enlightenment tradition. This attitude may be decried and a communal ideal of scientific practice held up, but it is problematic whether or not the ideal is achievable within the tradition by individuals adhering to the tradition. If there is, on the one hand, an individualistic, competitive system with which the scientist identifies and within which he or she exists, and, on the other hand, an ideal of scientific practice in which knowledge is pursued for its own sake and the fruits of work are available to all, then there are different milieus in which an individual scientist realises his or her identity. This may tend to produce a confusion of identity and self or, perhaps typically, the ability, on occasion, to hold and advance each view without undue discomfort from the incoherence involved.

The society in which the tradition of individualism holds and within which the scientists practice will be such as to encourage scientists to guard their work and seek to protect it through trademark, patent law and the law of intellectual property. Scientific work will be encouraged which is of benefit to the individual and to powerful groups. This leads to scientists being seduced by the prospect of power, riches and fame. Experiments and their results suffer. The interpretation of experimental results may become biased.

> Publish or perish used to be the mantra by which the researcher lived or died. Today, according to an increasing number of eminent scientists in unpopular fields, Australian researchers can do both.
>
> Unpopular here refers to any of those fields of science liable to produce evidence unsettling to the fixed world-view held by governments, business, special interest lobbies or that most anonymous and unaccountable of research controllers, the stakeholders. Nowadays it refers to environmental science, forestry research, greenhouse science, fisheries research, energy research, water research, atmospheric research, pharmaceutical research, dietary research, some health research and, in general, public good research. (Cribb, 2006)

This is not a problem peculiar to Australia.

In the practice of science the ideal is one of a society in which dissent is encouraged, criticism is expected and valued. It is a society in which the first move is always to attempt to find the weakness in another's argument. This inevitably leads to a community which is riven by dissension. Scientists are keenly aware of their rights and seek to enforce them. The purpose of science will be to predict and to control the physical world so as to shape it to ends that are desired.

In response it may be said that this paints an extreme picture of both the tradition and its effects and of science within that tradition. If it is a picture of Western liberal democracy and the practice of science within it then it is mistaken. There may be instances of science being practised as painted but they are aberrations. Scientific practice is on a road to an ideal state. However, it can be argued that such mishaps, and such scientific aberrations, are embedded in liberal democratic assumptions, embedded in the tradition. Western liberalism in placing such an emphasis on individual autonomy, choice, interests and achievement, gives opportunity to those who wish to take advantage of their fellows, it devalues community and social harmony. The scientific community is a reflection of the larger community. The response to this has been the claim that out of criticism, dissension and argument comes knowledge.

Consider now what it may be to educate in an individualistic tradition. Education will tend to be defined in terms of the initiation and development of the individual in the manner of Peters (1966). Watt (1989) argues that the analytic philosophy of education which has been so influential since the 1960s tends to exhibit individualistic assumptions. Curricula will be designed and written on the basis that the aim of the curriculum is the development of the individual as distinct from any social purpose or benefit. Teachers will be trained learning and using social science that has been developed within the tradition. For instance Piagetian psychology has been the mainstay of educational psychology courses in teacher training in England and Australia for the last 40 years. Piaget is a psychological researcher who is representative of the individualistic tradition. The emphasis is on the individual child or student exploring his or her environment and constructing his or her own individual account of it. As Watt points out:

> He consistently minimises the importance of what children learn from adults whether formally or informally. This is an inevitable result of the heavy emphasis which he puts on what children learn as an outcome of their own self-directed exploratory and problem-solving activity. (Watt, 1989, p. 176)

But this is acceptable within an individualistic tradition. However, it means that, given the attention accorded to the work of Piaget in the training of teachers, the practice of education tends to be limited by its disregard of the social environment within which children and students exist.

Teachers and students in the same individualistic milieu will also be affected. Teachers will be expected to see students as individual competitors and the curriculum organised in terms of the enhancement of the individual. Group work, which is commonplace at the present time in many Western countries will create problems for individual competitors who are placed in a communal situation and expected to work amicably together. Teachers and students will be encouraged to see the group as coming together for the mutual advantage of the members and a situation in which self-interest is the reason for working in the group. As a result, where, as typically happens, two members of a group of six do all the work there will be recrimination. Those that have done the work will see themselves as unfairly disadvantaged by the others and will demand a situation in which they receive the credit for the work at the expense of the others. Teachers will tend to accept this as a fair response and reproach those who did not contribute. In general, it can be argued that the communal situation of the classroom and the school in a Western liberal state is, necessarily, a place in which there is tension between the accepted individualistic mores of society and the need for some form of communal endeavour.

THE COMMUNITY VIEW

Consider now the opposing point of view which emphasises the community and the culture within which that community exists. For any given individual, membership of such a community or culture is not necessarily chosen and their social circumstances are not necessarily chosen. The viewpoint is that we are born into a culture and a community with all our social attachments and circumstances contingent on this. To a great extent, self-identity is constituted by this circumstance. As a result individuals cannot be understood or appraised *in vacuo* but must always be seen against the culture and the community from which they came and to which they owe their initial identity. The Enlightenment picture of the individual who stands outside experience and makes a rational appraisal is a myth. The individualism of the Enlightenment is a figment of the imagination. Individuals do not exist in a social vacuum. They take their identity in no small measure from their circumstances. Where the rights-based culture treats as fundamental the rights of the individual it thereby seeks to:

> … deny the same status to a principle of belonging or obligation, that is a principle which states our obligations as men to belong to or sustain society, or a society of a certain type, or to obey authority or an authority of a certain type. (Taylor, 1992, p. 30)

Over the last 30 years there has been a challenge to the place occupied by Western liberalism as an ideal. In the 1970s and 1980s it came from the political philosophy known as Communitarianism. In the 1990s a challenge on similar ground has come principally from East Asia.

> The deeper challenge to Western liberal democracy has emerged from the East Asian region. In the 1990s, the debate revolved around the notion of Asian values, a term devised by several Asian officials and their supporters for the purpose of challenging Western-style civil and political freedoms. Asians, they claim, place special emphasis upon family and social harmony, with the implication that those in the chaotic and crumbling societies of the West should think twice about intervening in Asia for the sake of promoting human rights and democracy. (Bell, 2001)

It is particularly appropriate to consider the East Asian challenge in the circumstances of the modern world and, particularly appropriate for other countries in Asia-Pacific, such as Australia and New Zealand, to consider that challenge. In doing so the issues about the translation of alien traditions, incommensurability and evaluation will surface again.

The countries of East Asia, Singapore, Korea, Japan, Taiwan, and China, have embedded in their cultures elements of Confucianism (Little & Reed, 1989; Smith, 1996; Tu Wei-Ming, 1996; Akihiro Chiba, 2004; Guo Xiaoping & Song Enrong,

2004). In saying anything about the presence of Confucianism in East Asian countries there is a presumption that it is possible to give a credible statement in English of the nature of the Confucianism which is embedded in these cultures such that it can be understood by those who have matured in another culture, Western individualism. This is the problem of translation of meaning.

The problem has produced a number of responses, one of which has already been mentioned. Radical incommensurability is the response which asserts that the concepts and world view of another culture may be such that it is not possible to give a meaningful translation into the framework of another culture. This view is posited on the supposition that concepts and world views may be so embedded in a culture that to attempt to express them in the absence of that context is to mistranslate. It appears to me that the empirical evidence is against this view. People do move between cultures and act in ways which indicate that they understand both cultures. They are able to express themselves about each culture. The translations of the Confucian texts into English are accepted as effective. In which case, concepts peculiar to Confucianism receive meaningful expression in English. The fact that it is possible to talk sensibly about good and bad translations presupposes that there are criteria which are usable and that these criteria do not simply refer to literal translations. All this is not say that there might not be some concepts which are untranslatable. In which case to understand them a person must become bilingual or bicultural. However all these remarks rest on a supposition. Confucianism is not merely a collection of sayings, injunctions and advice; it has a particular view of the world. It is from the base of a particular view of the world that Confucianism, like other cultures, develops particular concepts and approaches, a conceptual framework. If the base cannot be translated then the conceptual framework relying on that base lacks meaning to the outsider. To the extent that the base is only partially translatable then this affects the ability to translate the conceptual framework.

Guo Xiaoping and Song Enrong, of the China National Institute for Educational Research, Beijing, claim that the difference between Chinese and Western cultures can be seen:

> The difference in backgrounds and values between Chinese and western cultures with regard to relationships between humans and nature, is reflected in categorically different systems of thinking. While western culture emphasizes the need to prevail over and conquer nature, Chinese culture emphasizes the harmonious integration of people and nature; hence the theory of ‘heaven and man in one’. Today, when the ecological environment is deteriorating and the crisis of human survival is looming, the theory of ‘heaven and man in one’ deserves more careful study.
>
> This philosophical doctrine of ‘heaven and man in one’ in traditional Chinese culture has influenced the attitudes of Chinese towards nature for thousands of years. How to co-exist in harmony with nature is, even today, still a major concern of modern educators and philosophers. (Guo Xiaoping & Song Enrong, 2004)

This concept is picked up by Wong (2001) when he states:

> Daoism and Confucianism, at least after a certain stage in the development of these schools, exemplify the way that a set of interests intertwine with beliefs about the world. Both schools exemplify in different ways a conception of understanding the world that is inseparable from the interest in coming into "attunement" with it, to use a felicitous word from Charles Taylor (1982). To become attuned to the world is to see its goodness and to know one's place in the order of the world. (Wong, 2001)

'Attunement' appears to be part of the base on which a conceptual framework is erected. Radical incommensurability would be the view that 'attunement' cannot be translated. Taylor (1982) would take the view that it can be translated successfully. There are middle positions between these two extremes, Wong (2001) comments:

> Samuel Fleischacker (1992) proposes a more moderate version of incommensurability – sometimes we can understand others just well enough to know that we don't understand them. His argument has roots in Wittgenstein's view that knowledge depends on a background of shared assumptions and standards of evidence. "World pictures" are embedded within cultures. (Wong, 2001)

Radical incommensurability has already been attacked as incoherent, consequently, the choice is between something akin to moderate incommensurability and complete ability to translate. Bicultural people are of little use in solving this problem. They may understand each culture perfectly, but that is not the same as being able to translate the one into the other. However, they do give an indication that a view towards Taylor's end of the spectrum may be reasonable.

The problem was stated earlier – what is demanded of any culture or paradigm is that it is sufficiently characterised for it to be possible to identify the claims that are being made? What is necessary if any culture or paradigm is to be seen as incompatible with another is that its claims about a topic are given sufficient expression for it to be clear that the claims being made are incompatible with the claims of the other culture about the same topic. If the view is taken that 'attunement' is a fundamental response to the question as to what counts as human flourishing then, the problem is to give sufficient expression to that concept so that the claims that are implicit and are derived from it can be assessed as incompatible or otherwise with the claims made by, for instance, Western individualism, about the same topic. It is then a further question as to whether or not an assessment can be made as to which response to the question of what counts as human flourishing is to be preferred. If an assessment is to be attempted then the first step is to examine the conceptual framework for internal coherence. The second step is to assess whether or not it deals with human flourishing better than its rival. In using the word 'better' the implication is that both views may have merits.

It is beyond the scope of this book to investigate this issue. I shall follow Taylor (1992) and take the view that it is possible to speak meaningfully about

Confucianism as it is embedded in today's East Asian culture. Consequently, it is possible to consider the attitudes and virtues which are embedded in East Asian culture as a result of Confucianism.

An inquiry into East Asian industrialisation and the part played in that by Confucianism is introduced by comments such as:

> The classic Confucian vision that "only when families are regulated are states governed" (stated in the opening passage of the *Great Learning*) is still taken absolutely seriously in East Asian political culture. (Tu Wei-Ming, 1996, p. 8)

The notion that the state is the family writ large is fundamental to Confucianism. The authors who contribute to the volume all trace various ways in which Confucianism has played and is playing a significant role in East Asia. Liu Shu-hsien (1996) argues that the reaction of Chinese thought to the impact of Western thinking and, in particular, its science and technology can be traced as three overlapping stages:

- shock at the superiority of Western science and technology;
- a feeling that their culture was a barrier to absorbing Western ideas and the immediate need to do so; and
- the realisation that Western culture was not completely positive and their own culture had a lot to offer.

Liu Shu-hsien argues that China, and by implication the other East Asian countries, are currently in the third stage. The third stage will clearly promote a fresh look at the Confucian heritage of the country and a re-affirmation of the worth of some of its values in modern society. Smith, writing on the Japanese family, lists typical Confucian virtues and comments:

> What moral code, what ethical teaching, what set of principles (dare we ask, what religion) fails to extol these virtues? If, on the one hand, we assert that they are specifically, if not exclusively, Confucian, then there can be no doubt that the Japanese family in particular and Japanese society in general are Confucian as well. If, on the other hand, we mean only to say that insofar as the family that looms large in the lives of most Japanese is conceived to be the sine qua non for a stable society, then I think it equally the case that the Japanese family and Japanese society are Confucian still. (Smith, 1996, p. 174)

Confucianism has as an ideal the moral transformation and self-realisation of the individual within society.

> In this ideal a social commitment is balanced by a self-regarding commitment to the moral fulfilment of the human individual …. The orthodox Neo-Confucian commentaries on the Four Books, such as Chu Hsi's, made it clear that Confucianism envisioned two goals: the moral perfection of the individual self as well as that of society. The relation between the two goals

> is one of interdependence rather a one-way passage in either direction. (Chang Hao, 1996, p. 74)

The morally perfected individual is the virtuous individual, a lover of virtue. This is so whether the individual is a peasant or a ruler. Confucius linked virtue to knowledge in a manner similar to Socrates. So virtue is the inevitable product of knowledge and vice the product of ignorance. There is here a clear connection between Confucianism and the necessity of a sound and rigorous education.

> Confucianism is a perfectionist virtue ethic, if such an ethic is distinguished by its central focus on three subjects, character traits identified as virtues; the good and worthwhile life; and particularist modes of ethical reasoning. These three subjects are interrelated. The virtues are traits of character necessary for living a good life. The virtues typically involve acting on particularist modes of ethical reasoning that do not rely on deducing specific action-guiding conclusions about how to act from general principles but rather on judging in the context at hand what needs to be done. (Wong, 2001)

There are clear similarities with an important vein of thought in the Western tradition. In particular there is an obvious similarity with the work of Aristotle. More recently there is similarity with the work of Foot (1978c), Hursthouse (1999), McDowell (1995) and MacIntyre (1981, 1988, 1990a, 1999). Aristotelian and modern virtue ethics start from a consideration of human flourishing and argue for the exercise of the virtues in contexts which are particular. Moral decisions demand *phronesis*, practical wisdom. However, as Wong (2001) notes there are differences between Confucianism and Aristotelianism. The major differences being the central importance of family and a demand that the virtues be practised aesthetically in Confucianism.

Modern Confucianism emphasises, as did Confucius, four virtues as being fundamental to the morally perfected life –compassion or benevolence, sincerity, filial piety and propriety.

Benevolence was central to the thought of Confucius. This virtue consisted of two parts, the method of determining what is benevolent, *shu*, on this occasion, and the implementation of that determination, *chung*:

> The essential point about benevolence is to be found in Confucius' answer to Chung-kung:
>
> Do not impose on others what you yourself do not desire. (XII.2) (Lau, 1979, p. 15)

and this is followed by:

> As the way of the Master consists of *chung* and *shu*, in *chung* we have the other component of benevolence. *Chung* is the doing of one's best and it is through *chung* that one puts into effect what one had found out by the method *shu*. (Lau, 1979, p. 16)

Benevolence encompasses a regard for the welfare of others and an acceptance of obligation to others in need. It is viewed as a crucially important social virtue:

> Confucius' social philosophy largely revolves around the concept of *ren*, "compassion" or "loving others." Cultivating or practicing such concern for others involved deprecating oneself. This meant being sure to avoid artful speech or an ingratiating manner that would create a false impression and lead to self-aggrandizement. (*Lunyu* 1.3) Those who have cultivated *ren* are, on the contrary, "simple in manner and slow of speech." (*Lunyu* 13 27). For Confucius, such concern for others is demonstrated through the practice of forms of the Golden Rule: "What you do not wish for yourself, do not do to others;" "Since you yourself desire standing then help others achieve it, since you yourself desire success then help others attain it." (*Lunyu* 12.2, 6.30). (Riegel, 2002)
>
> Benevolence is linked to another major virtue, reverence (*ching*).
>
> *Ching* … is born of the awareness of the immensity of one's responsibility to promote the welfare of the common people. (Lau, 1979, p. 26)

Sincerity encompasses truthfulness, promise-keeping, conscientiousness in dealing with others and the acceptance and discharge of obligations. Filial piety represents the duty that children owe to their parents to love and respect them and to bring them honour by the way in which they lead their lives. Propriety may be summed up as being the virtue of doing the right thing at the right time. For instance, an individual should repay injury by seeking justice, but return kindness with kindness.

The modern practice of Confucianism flows from the basic emphases above. Confucianism emphasises social order and social harmony derived from human virtue rather than the law. There is high value placed on hard work, education, self-discipline, the family and the community. It is a view that emphasises the responsibilities of the individual to the community rather than their rights against that community. It is a view which emphasises that the interests of the community take precedence over the interests of individual.

Japan received Confucianism which became intermingled with Buddhism and Shintoism throughout the course of Japanese history. Values, which are now properly called traditional, emerged from this mix.

> … many of the traditional values attributed to the Japanese today originate from this heritage. Maintaining social harmony, respect and obedience of the elderly, loyalty to the superior, avoiding arguments and disputes, maintaining peace and harmony, social behavior with modesty and etiquette reflecting the social status, care and compassion, and the importance of deed rather than promise, are all important values and the virtues which sustained the feudal system. In a neutral sense, they may be considered positive values even today for living in a modern democratic society. (Akihiro Chiba, 2004)

Akihiro Chiba then goes on to argue that the traditional values still play a part in modern Japan and that there is a tension due to the conflict with Western values which have entered the country since the end of World War II.

> The traditional values of the Japanese such as diligence, hard work, modesty, humility, respect for order, obedience, perseverance and economy all contributed to re-establishing Japan as a peaceful nation, and also to national reconstruction and development. In fact, Japan has achieved a spectacular performance in its post-war recovery and its economic and technological development. Japanese people have a strong characteristic of collective being rather than individual existence which manifests in a tremendous power to mobilize and unite at times of national crisis or emergency, but they often have been criticized for lacking the qualities of individuality and self assertion that are basic values of democracy. Uniformity and conformity are often given preference over uniqueness of individual personality, thinking and behavior. (Akihiro Chiba, 2004)

It is clear that Japan and China share a range of traditional values and that these values are embedded in the culture. Other countries, such as Singapore, Taiwan and Korea share this base of traditional values and it is in this sense that it is possible to talk of East Asian values.

As might be expected, social and economic concerns take precedence over civil and political concerns. Social harmony is valued above dissent at both the group level and the national level. The notion of harmony also extends to the view of the world as a whole. Ideas, theories, beliefs are expected to harmonise. Harmony in Confucianism is harmony between Heaven and human beings, between human beings and nature and between human beings. Discord, whether in the life of a person, in the affairs of a community or a state, in the understanding of the world of our existence, is to be avoided or to be resolved.

There is clear consensus amongst internal and external commentators that Confucianism is embedded in East Asian culture. There is a cultural reappraisal taking place at the moment and it would seem that East Asian society is seeking ways to evaluate Western culture and absorb that which is found to be valuable. The important point is that the evaluation will take place on the basis of criteria which are partly Confucian. It is perhaps an East Asian version of the old Irish advice to emigrants – 'integrate but never assimilate'.

The virtues of Confucianism are seen as social values and intrinsic to the life of the community and the individual within it. This is a teleological view of the world. The *telos* of the individual is to achieve the highest realisation of his or her capacities and this can only be through their participation in education, the community and culture in which they exist and to which they owe obligations. Confucianism is an invitation that commences with questions about the proper end of a human being, with questions about human flourishing.

This view of the world can give rise to a number of social forces, pressures and arrangements as a society attempts to realise Confucianism within its particular circumstances. It is possible to identify examples of these forces, pressures and arrangements which tend to be typical in East Asia. Little and Reed give an account of what they term 'Confucian Characteristics' (Little & Reed, 1989, pp. 54-55). The characteristics are a mixture of ontological stances and social forces, arrangements and pressures. It is an account of a Confucian society as one which has several characteristics:

(a) an inter-active web of social pressures to ensure that all members of the society accept their obligations and responsibilities and share the rewards of that society;

(b) a system where competition is clothed in ceremony and ritual;

(c) a system where dealings between individuals and groups takes place in such a way as to preserve social harmony and consensus;

(d) an education system that is demanding of students;

(e) an education system which is highly competitive;

(f) an awareness of its historical narrative;

(g) is less interested in scientific breakthroughs than in science of value to the community; and

(h) an acceptance of authoritative government.

And it is this last item that frequently grounds Western liberal criticism of East Asian practices. It is useful to look briefly at the concept of authority. It is necessary to make a distinction between having *de jure* authority and having *de facto* authority, followed by a distinction between being 'in authority' and being 'an authority'.

De facto authority may be defined as the authority that a person has and is to be seen in the degree to which others obey his or her commands. *De jure* authority is that authority conferred on a person or persons by a system of rules and these same rules will also place an obligation on others to obey that person. Clearly the situation may be such that he or she may have less or more authority than he or she is entitled to. Authority in the above sense is clearly distinct from the possession or exercise of power, for power does not rest on any rule system but on the ability to subject others to one's will. Authority may of course be supported by power but this does not justify confusing the two.

> The concept of authority is inseparably connected with a rule-governed form of life. It is only appropriately applied when there is a question of something thought, said, or done being correct or true. It is inseparable from the use of speech, symbolic gestures, and rituals by means of which verdicts, commands, and pronouncements, which lay down what is correct or true, are enunciated and vested with significance. (Peters, 1966, p. 238)

We speak of people being 'in authority' and being 'an authority'. To be an authority is to be someone who is recognised as having a level of knowledge in a particular area such that their opinion is entitled to be respected in that area. The concept implies that society has set up some system whereby it is possible to recognise authorities in this sense. Within the practice of science the community of scientists give recognition to such authority figures. Authority figures recognised within a practice may also be recognised by society at large as authorities. Within any practice there are criteria by which authority figures are recognised. In science it may depend upon publications, degrees or discoveries or a combination of such factors. What is important is the recognition by the community of practitioners. The important point to note is that authority in this guise depends ultimately on the reasons given for the pronouncement. The justification for the respect given to the pronouncement does not lie, ultimately, in the person but in the reasons that the person is able to put forward to support what he or she says. In science, as in other practices, the reasons are evaluated according to the criteria of the practice.

To be in authority is to be in a position where you may require a person to obey on the grounds that a system of rules gives you the entitlement to issue the command and obliges the person to obey. Society is replete with authority figures who are entitled to command obedience from citizens. Police are an obvious example. The justification for such authority lies in the consent of society to its grant and to the system of rules that enable the grant. The justification for any command does not lie in the reasons that may support it but in the office of the person in authority and the rules which specified the office and the limits of its authority.

Nothing in the above enables a distinction to be made between Western liberalism and Confucianism. In both cases authority rests on a system of rules and values. The difference between the two traditions lies in the extent to which and the manner in which authority is respected and questioned. It could be argued that in Confucianism there is a respect for authority founded on a demand that those in authority should be the best placed to exercise authority. That is, society places in authority those who are an authority. This is in contrast to Western society which does not see the need always to place in authority those who are an authority. The growth of managers whose expertise is in managing is an example. The election of politicians of no particular knowledge is another. It would seem arguable that there is a demand for educational achievement in those societies which have their cultural roots in Confucianism which connects to the respect for those who are authorities and connects to the grant of political and civil authority to those who are authorities.

Part of the challenge from East Asia has been to identify the science of the West as a specific form of Western science imbued with values and shaped by a tradition which is peculiar to the Enlightenment. Western science is seen as marked and corrupted by the baggage that it carries from the tradition within which it is formed. East Asian society talks of the need to avoid spiritual corruption when it considers the impact of Western liberalism. It is an ontological and epistemological challenge. It is also a moral challenge. The replies of Western liberalism are fairly obvious. There would be argument that civil and political rights are paramount and that to give them away to authoritarian government is to demean the individual.

The response to that would be that social and economic rights come first because first there must be a fulfilment of obligation to the community. There must be bread on the table before there can be argument about freedom. The response to the charge of authoritarianism is that the Western liberal individual mistakes the nature of authority. In some ways Western individualism and Confucianism are complementary. Perhaps the solution is to see them as complimentary, the Yin and Yang?

It is arguable that the individual in the cultures in which Confucianism is embedded is not understood as complete apart from the social order within which he or she exists. That is, an individual's identity is moulded and defined by the social order and this is in contrast to the Western liberal world.

> We find instead [of western individualism] in Japanese thought an entirely intelligible hospitality to, at the one extreme, doctrines such as those of Confucianism, in which the self is understood entirely in terms of social roles and responsibilities, and at the other Zen Buddhism for which the self is no-self, is in itself, so to speak, a nothing, to be pointed at only by paradoxes which make of that pointing a no-pointing. For, take the Japanese self, in its self-understanding, away from social roles, and what you have is a self that is not yet or no longer. It is no accident that there are no Japanese Aristotelians any more than there are Japanese Wittgensteinians. For the terms of Japanese philosophical debate, except insofar as they are affected by the incursions of the West, presuppose a conceptually structured tradition quite other than and alien to that presupposed by Western philosophy. (MacIntyre, 1990b, p. 495)

If MacIntyre is correct then it leads to the view that any successful stripping of the social roles from a Japanese is to run the risk of psychological damage which, in the extreme, may be irreparable.

In China the teaching of traditional values is considered an important part of the school curriculum.

> It is a major task for primary and secondary schools to teach about the traditional morality of the Chinese nation. Schools need to actively explore effective means of doing so, means that are adaptable to their own conditions and advantages. At present, in addition to classroom teaching and related extra-curricula activities, several schools compile textbooks of their own, availing themselves of local historical and cultural traditions to conduct moral education focusing on local values. (Guo Xiaoping & Song Enrong, 2004)

The actual curriculum clearly focuses on traditional Confucian and Chinese values.[2]

Consider group work whether within or outside school. There will be no necessary tension in the formation and work of the group deriving from the social tradition within which it is formed. Members of the group will tend to accept that

[2] See Appendix.

they work for the benefit of the group. This can mean that the individuals within the group see themselves as responsible to some degree for the learning of the other members of the group. The group is viewed as a cooperative in which the members see themselves giving of their particular ability and being given according to their needs by the others. The harmonious working of the group will be an aim of the members of the group.

Consider now what it may be to practise science in East Asia within a community that has its cultural roots in Confucianism. I accept that what I shall argue does not apply to the whole of Western science nor to the whole of East Asian science. There is, however, sufficient in both cultures to enable a significant comparison to be made and it is certainly true that there are many in East Asia who see the comparison as valid.

What might be said of a scientist practising in a Confucian tradition? The practice of science will tend to be orientated towards those areas which are beneficial to the community. Within the community of scientists value will be placed on working together in harmony. There will be a view that dissension can be counter-productive and criticism an act of aggression. The construction of theories and the interpretation of results will tend to favour a teleological approach. Scientific theories and explanations will tend to be examined for their harmonious relationships with other realms of belief and knowledge. The basic notion of harmony between Heaven, mankind and nature is paramount. Science will not be treated as a separate, individual discipline which may be pursued in isolation. Science will be taught and practised in a rigorous manner and, while there will be competition and criticism between practitioners, it will be controlled and channelled by ceremony, ritual, courtesy and what is of value to society and contributes to harmony.

There is clear contrast between such a view and the Western view. It is a contrast that surrounds a shared core. Just as human beings are common to both the Western community and the East Asian community so the notion of science as an empirical investigation of the world in which we live in the attempt to produce explanations of that world which are then checked against the world is common to both Western and East Asian science. But the practice of science is different, the value of science to the community is different, and the place of the community of scientists in each culture is different.

Bell (2001) writing about Communitarianism notes three arguments supporting the view that cultural factors may influence decisions on rights. Those arguments can be applied to the practice of science and its teaching and to them I add a fourth. Cultural factors most obviously come into play in situations where the choice is difficult and not clear but this is not to say that they not operative at other times.

– Cultural factors can affect the prioritising of scientific work. Given that there is a limited amount of time and resources available for scientific work there has to be some prioritising. This may involve difficult choices. Choices made against a Confucian background may, on occasion be different from those made from a Western liberal background. Medical or agricultural research may take precedence over basic physics. Technology becomes more important than basic research for technology is seen as more beneficial to the community. In education there are choices to be made and cultural factors will play a part in decisions. Decisions have to be made not just about the curriculum but also about the type of resources and their allocation.

- Cultural factors can affect the justification of the decision. Medical or agricultural research may be justified on the basis that it is more valuable to the community than basic physics. In education the justification of curriculum choices and the allocation of resources may be justified on Confucian grounds.
- Cultural factors can lead to the formation of distinctive practices and institutions. The Chinese have many universities but many of them are distinctive in their practical, community orientated aim. Their names may indicate this, witness the Chinese Pharmaceutical University.
- Confucianism, like Aristotelianism, is a teleological system. In addition there is always an attempt to harmonise science and nature. The practice of science in such a system will search for coherence and, in matters of choice, favour a teleological explanation that emphasises and promotes the harmony between humanity and the world. This is a cultural factor at work.

It can be said that there is a Western science and an East Asian science. But does that produce a commitment to a multiplicity of sciences? For instance, there are another three major world views in Asia which might well be seen to produce a particular view of science. They are Buddhism, Hinduism and Islam. On the basis of what I have argued there seems no obvious reason why it should not be that Buddhism, Hinduism and Islam produce views of science which reflect their world views. In which case it would be particularly important for the scientists, science teachers and curriculum planners of Western cultures in the Asia Pacific to take note of this; not to mention politicians, industrialists and others. But is this just the beginning of a long progression of different sciences which each reflect the cultural tradition in which they exist? To be consistent with the previous argument the answer must be in the affirmative. However, this is not to descend into cultural chaos. The practice of science in any culture must conform to what counts as science. However, that practice takes place within a cultural milieu and this will affect the practice. At times it will affect the theories and hypotheses developed by the practice. A tendency in East Asian science, in matters where there is a legitimate choice, to favour theories and hypotheses which are consistent with Confucian harmony would be an example. Note, however, that it has to be a matter of legitimate choice. In other words the standards of what counts as good science apply.

It is now evident that there is an important connection between the Aristotelian views that I have been arguing and Confucian views and, consequently, between both these world views and the practice and the teaching of science and the virtues.

CHAPTER 14

WHY THE VIRTUES ARE NECESSARY

The flourishing animal

The virtues have risen in the context of cultural science and they raise a point of view incompatible with the Humean distinction between 'is' and 'ought' and Western acceptance of that distinction.

It has been commonplace in the thinking on ethics from the time of Hume to assert that 'good' and other ethical terms are distinct from the rest of language. That is, there is language which describes and there is language which evaluates. This has led to an increasing gap between the consideration and description of the world of our experience and the consideration of the world of ethics. The 'is-ought' distinction is the manifestation of this gap.

For Aristotle and those that followed him until the time of Hume there was no such gap. For Aristotle, ethics had to be based in some way on a consideration of human nature and what counts as human flourishing. There have been doubts about whether such an enterprise could succeed and it has largely been dismissed in Western European philosophy as a feasible enterprise until recently (Hursthouse, 1999, pp. 1-2). The focus for the last 300 years has been on such projects as the Kantian categorical imperative, utilitarianism, emotivism and the like. Now in the hands of Anscombe, Foot, MacIntyre and others the Aristotelian approach is revived. The account of ethical naturalism that follows is influenced by Anscombe (1995), Foot (1978a, 1978b, 1978c, 1994, 1995), Hursthouse (1999), McDowell (1995) and MacIntyre (1981, 1988, 1990a, 1991, 1994, 1999), though none of them would necessarily agree with all that I say.

The word 'good' is an attributive adjective. We may talk of a good knife and we are referring to its ability to perform the function for which knives were created. That is a different attribution from remarking that this knife is a good one to give away because it is of no further use. The point being that criteria are embedded in the usage. In the same way we may talk of a good sheepdog and implicitly refer to criteria which are built around the characteristic function of a sheepdog. So much is straightforward. Foot's original point was that it appeared strange that we should suppose that this analysis no longer applies when we talk of a good human action, that is, when we talk of ethics.

MacIntyre has observed that holders of the 'is-ought' distinction must assume that no moral arguments involve functional concepts and that this involves the rejection of the Aristotelian tradition that all moral arguments have as a central idea the functional concept of human flourishing (MacIntyre, 1981, pp. 58-59). A good act is that which, characteristically, a good person would do and this is a factual statement about human flourishing.

Hursthouse (1999) follows Foot in arguing that a living thing such as a plant or animal is evaluated as a flourishing or good example of its type on the basis of two criteria.

- Does the plant or animal contribute to its individual survival so that it may have the characteristic life span of the species?
- Does the plant or animal contribute to the continuation of the species?

To these two criteria Hursthouse adds two more which she claims are peculiar to animals. They are derived from the recognition that plants 'react' whereas animals may be said to 'do' things. Each criterion is therefore qualified with respect to animals. Animals have to take measures to ensure that they survive. Hunting animals hunt and need to be able to hunt well if they are to survive. Many animals take measures to rear their offspring and teach them how to survive. The more complex the animal the more complex the measures the animal has to take in regard to acting for their individual survival and the continuation of the species.

Hursthouse continues her argument by asserting that:

> Somewhere along the ladder of nature – once again, quite possibly somewhere indeterminate – we start ascribing pain, and somewhere – perhaps at the same point, perhaps not – pleasure. Animals that can feel pain and that are capable of pleasure or enjoyment are evaluated in relation to not only the first two ends, but a third one, namely characteristic freedom from pain and characteristic pleasure or enjoyment. (Hursthouse, 1999, p. 199)

Hursthouse notes that a related point is made by Foot (1978c) with regard to animals benefiting from comfort and relief of suffering. For Hursthouse characteristic freedom from pain is a benefit but must be seen in relation to the feeling of pain being a survival mechanism. For a horse not to feel pain when a stone is trapped in the frog in its hoof is to be defective as a horse. Hursthouse, is now taking the matter further than Foot. I have no quarrel with the notion as she puts it. I do have doubts about her next move.

> Closely associated with the ascription of pain and pleasure to the more sophisticated animals is the ascription of a certain, at least minimal, psychology, of emotions and desires. Animals capable of such a psychology are evaluated as good or defective not only with respect to the three other aspects but also with respect to some emotions and desires. (Hursthouse, 1999, p. 200)

Hursthouse then moves up, as she puts it, another rung of nature and considers the social animals to whom she ascribes the end of good functioning of the social group. She notes that bees and ants cause a problem for the smooth progression up the rungs to social animals for we do not ascribe pain and enjoyment to bees and ants.

Can a characteristic dog feel the emotion of fear? I feel fear when confronted by a brown snake. Why? Because I recognise the danger to myself and my probable inability to control the situation effectively. A baby, who had not learnt that snakes are dangerous would not feel fear. Dogs may be programmed to recognise danger or they may be taught, of this I am unsure. Certainly dogs appear able to remember and on this basis may display physical signs characteristic of an emotion when confronted with a present situation that brings forth a memory of the past. The emotion is a response to a view of the past seen in the context of the present. Looking back on some events in my life, the birth of my children for instance, brings an emotional response. I can recall and contemplate, reflect on the past quite deliberately. Are dogs able to do this? Some would argue that dogs cannot look back in this sense and therefore cannot experience that emotional response. I am unsure of this.

Consider this. I am on sentry duty in the front line. The night is dark and the enemy is known to send out patrols to attack and kill sentries. I feel fear not because of anything that I can see or sense but because I can reflect on the possible future, I can imagine what might happen. Some would argue that animals such as dogs or apes or dolphins cannot consider the future and thus are incapable of feeling fear where a consideration of the possible future is involved. But it appears that animals other than ourselves can sense a future danger. There may be a doubt that animals, apart from ourselves, have the ability to consider the past and the future. The temptation is to confine the capacity to reason, to reflect, to ponder, to contemplate, to speculate and to imagine to ourselves. Hursthouse in putting forward a criterion based on emotion and desire misses the point. The point being that emotion and desire are based on a cognitive core. Without that cognitive core there is no emotion. There may be physiological states but that is a different matter. Hursthouse does not delve far enough into emotion and desire, she leaves it at the level of psychological and physiological states. At the same time, and perhaps consistently, Hursthouse appears committed to the commonly held position that the capacity to reason separates us from the rest of animal kingdom. In my view to do so is to forget Darwin. Evolution is a process by which species change takes place and the process is essentially a slow one without radical cut-off points. The message of Darwin is that we have evolved and therefore it is unreasonable to separate ourselves from the other animals by postulating a radical and unique cut-off. We are different in degree not in kind.

It appears from palaeoanthropology that human beings have the same ancestral line as chimpanzees and that human beings are between 98 and 99 per cent genetically similar to chimpanzees. Perhaps even more interestingly it appears that human beings are very similar to each other genetically, certainly more so than chimpanzees are to each other. That ancestral line from chimpanzees gives rise to a variety of creatures which have been classified as hominids. The class has been broken down into a variety of sub-classes all of which have given rise to argument and opinion. Claims are made to distinguish *homo ergaster, homo neanderthalensis, homo rudolfensis, homo heidelbergensis, homo erectus* and *homo antecessor* with the final sub-class being *homo sapiens*. The classification arguments result from the paucity of hominid ancestral bones, which is strange given the numbers of bones of other animals that have been found. Whatever view is taken of the evolution of *homo sapiens* from chimpanzees, it is a mistake to regard it as a progression from rung to rung. The progress is not linear but is best regarded as involving a number of side tracks and branches. The tendency to think of ourselves as the culmination of a linear, rational process dies hard, perhaps for reasons of human vanity (Tattersall, 2002). To think in terms of side tracks and branches introduces the probability of a number of dead ends until finally, possibly from a small population given the similarity between all human genetic makeup, *homo sapiens* came into being. In the light of this it is unreasonable to think in terms of a distinct and sharp break in the evolutionary pattern. The capacity to reason is a capacity that exists on a scale from the simple to the complex. We are able to observe the development of that capacity in the human infant and child, initially by observing the actions of the infant, and later, by observing actions and listening to their speech. At this point we can either take the view that the capacity is denied to other animals, particularly our ancestral hominids or we can admit that our ancestral hominids had an evolving capacity to reason. It is clear from the actions of chimpanzees and other apes that they perform actions which, in the human infant and child, we would class as exhibiting reason. It is clear that chimpanzees and other apes can communicate with each other, as can dolphins. Consequently, it is reasonable to take the view that the capacity to reason developed with the evolution of the hominids until we reach *homo sapiens*. It is unreasonable to think in terms of our genetic ancestors being unable to reason. Similarly, it is unreasonable to deny that other animals can reason.

I accept the line of the analysis that sets out the possibilities of a functional argument in relation to living things and the ascription of 'good' or 'flourishing' to them. Then, as Foot and Hursthouse have said, why should we suddenly change the basis of our ascription of 'good' or 'flourishing' when we come to human beings? If, as Foot and Hursthouse have done, an argument is run which is a teleological assessment of animals, then such an argument is underpinned by a notion of evolution. It then becomes unreasonable to refuse to continue that underpinning when it comes to human beings. We are animals, something which philosophers and others tend to forget. We are a different species that has evolved and, as with other species, it may be that we have to add criteria for the ascription of 'good' or 'flourishing' to ourselves, but this is not to say that the whole basis of the ascription changes. The difference between ourselves and other animals is one of degree not of kind. The argument remains teleological. Hursthouse makes the same point when saying that her argument may be incorrect in substance but remains correct in structure.

It is important to notice that the assertion that this is a good or flourishing gorilla is an objective statement. It does not depend upon the wishes, wants or desires of the individual but depends upon the evaluation of the gorilla against particular criteria. These criteria set out what counts as a characteristically good or flourishing gorilla in terms of its capacity to survive, to continue the species, to be a functioning member of the social group; a necessary but not sufficient list. If the particular gorilla satisfies the criteria then it is a good or flourishing gorilla. But this is not to say that it is an excellent gorilla. The satisfaction of the criteria is on a continuum where it usually has to be said that this gorilla is, for the most part, a flourishing gorilla. This particular gorilla might, for instance, be deficient in the art of grooming but excellent with the young or as leader of the pack. As in biology generally, it is rare to get a perfect example of a species, generally it is an evaluation that this individual satisfies the criteria for the most part. This raises issues of biological categorisation. Gorillas are distinguished from other apes on the basis of certain criteria. Biologists could conceivably change the criteria in such a way that gorillas are put in the same category as monkeys rather than apes. This is not, however, a stumbling block to the argument. The argument depends upon an estimation of the characteristic gorilla. However we might categorise them biologically, it is the case that bands of what we call gorillas do live together in social groups and can be examined as to their characteristic life together. The ascription of good or flourishing does not demand perfection. Because of this it will be the case that, while on many occasions we can say with conviction that this is a good gorilla or a bad gorilla, on some occasions we may be in an indeterminate position. This is to be expected for individuals differ but we should expect the majority of the species to be of the good variety for the continuation of the species.

The flourishing human being

To return now to Foot's basic query. If all this makes sense with respect to animals, and remembering that we are animals too, why should not the same type of argument apply to the ascription of flourishing to human beings? Clearly in respect of animality the argument may be run, but do our differences from the other animals make a difference?

The argument with respect to living things other than human asks what it is that a characteristic member of the species is or does. Therefore, the first point to make with respect to human beings is that they are social animals and, consequently, what counts as a flourishing human being will be one that has the capacity to survive for the normal life span, is able to continue the species and is a functioning member of the social group. On these counts someone with a terminal illness is a defective member of the species, someone who is infertile likewise and so is the isolate. Such a person is not a flourishing human being on the basis of those criteria. At this point Hursthouse suggests that we tend to separate our evaluations about our health from our evaluations about ethics. She suggests that we leave the health evaluations to biology and medicine (Hursthouse, 1999, pp. 206-207). I have doubts about this move. With animals, including ourselves, being healthy specimen is important to enable survival and the continuation of the species. It is also important for social animals for a certain measure of health is required to participate in and perform a variety of social tasks. There is, in my view, no need to make a break between ourselves and the rest of the animal kingdom at this point. This would be to make the same mistake that Foot and Hursthouse criticise. What can be emphasised is that some parameters of health are less important and some more important with us than with other animals because of the differences between us and them. This is as a result of our ability to organise some of our societies to minimise the effects of certain types of ill-health. For Hursthouse to separate what she terms the merely physical from the rest is in my view a mistake and undermines the unity she claims for the rest. MacIntyre (1999, p. 73) argues that we are all disabled to some degree and exist on a scale of disability. He continues:

> For it is and perhaps always has been a common assumption that blindness, deafness, deformed or injured limbs, and the like exclude the sufferer from more than a very, very limited set of possibilities. And this has often been treated as if it were a fact of nature. What is thereby obscured is the extent to which whether and how far the obstacles presented by those afflictions can be overcome or circumvented depends not only on the resources of the disabled – and these will vary a great deal from individual to individual – but also on what others contribute, others whose failures may be failures of imagination with respect to future possibilities What disability amounts to, that is, depends not just on the disabled individual, but on the groups of which that individual is a member. (MacIntyre, 1999, p. 75)

The human recognition of disabilities and reaction to and treatment of those disabilities has changed over the course of human history. History provides us with numerous examples of a reaction to disability which ostracises the sufferers or removes them bodily from human society. The European asylums of former times are called to mind as is the attitude to leprosy. In extreme cases the sufferer is put to death. Disability was often thought to be a judgement on the sufferer or the presence of the devil. In most modern societies disability is more readily seen as a physical or mental limitation which can be ameliorated and may not affect other human capacities and should not exclude the sufferer from human society. There is in this change a clear linkage between biological and medical knowledge and human capacities. The attribution or not of flourishing follows. Some cultures seem to have become less prone to failures of the imagination.

It is a further stage in our thinking on the matter to recognise that we are all on a scale of disability. To assert a gap between biology, medicine and ethics is to make the same category of mistake that is made when the attribution of flourishing is supposed to change as soon as we consider human beings.

The next issue is the difference between ourselves and the rest of the living world, in particular those animals closest to us in capacity. Foot (1995, p.9) provided the initial structure which Hursthouse develops and they share the commonly held view that the capacity to reason is the distinguishing mark of the human animal.

> Our single characteristic 'way of going on' is in a rational way, i.e. in any ways we can rightly see as something we have reason to do. Ethical evaluations cannot be a branch of biology or ethology because neither we, nor our concepts of 'a good human being' and 'living well as a human being' are completely constrained by nature. With the other animals, it is (almost) guaranteed that overall, there are good, healthy members of the species, who, with a bit of luck, will be living well, for what counts as being good and living well is determined by the standard that, so to speak, nature has laid down for them. (Hursthouse, 1999, p. 228)

McDowell, after imagining a rational wolf, comes to the conclusion that:

> Reason does not just open our eyes to our nature, as members of the animal species we belong to; it also enables and even obliges us to step back from it, in a way that puts its bearing on our practical problems into question.
>
> With the onset of reason, then, the nature of the species abdicates from a previously unquestionable authority over the behaviour of the individual animal. (McDowell, 1995, p. 154)

McDowell then develops this argument in the attempt to show that with reason we develop what he calls a 'second nature', that is, the dictate of the virtues.

Hursthouse is quite explicit about her adoption of the idea that reason is a distinctive mark of the human species:

> The third thing I import from Aristotle is a pair of interrelated distinctions.
>
> (1) There is the distinction between acting from reason, which we, typically, do, and what the other animals and small children do when they 'act'.
>
> (2) There is a distinction between rational wanting or desire, which we, typically, have and the mere passion or desire that impels the other animals and small children. (Hursthouse, 1999, p. 14)

But she does not see this as a hard and fast distinction but rather that there is a continuum between ourselves and the other animals with the young human child being an animal who has to grow into a rational human being who has, to some extent, put to one side animal and childish impulses and desires. However, it seems to me that she does not accept that animals can and do reason and this may be the result of confusion between reason and reflection. The notion of reason is notoriously hard to pin down to one particular meaning or use (Warnock, 1967, p.83) but this need not be a stumbling block. McDowell (1995) had within his argument an idea that is worth pursuing.

CHAPTER 15

REFLECTION

Foot, Hursthouse and McDowell are mistaken if they see reason as the capacity that divides humans and the rest of the animal world. Our knowledge of the ways in which animals, particularly the primates, act, leads us to the conclusion that they can reason. If we were to view such behaviour in human beings we would ascribe reason to them. Why should we not do the same with animals or must we persist in repeating the mistake that we made with the attribution of 'good'? However nothing hangs on whether or not animals can reason if the argument is that it is reflection not reason that is the crucial capacity that is the distinctive human way of going about things.

Consider the phrase 'rational reflection', a phrase that McDowell uses. The implication is that reflection is different from reason. McDowell considers the case of a rational wolf, that is, a wolf who:

> … would be able to let his mind roam over possibilities of behaviour other than what comes naturally to wolves. (McDowell, 1995, p. 152)

However, later, in describing the rational wolf, faced with the question of what he should do, McDowell's wolf becomes the 'reflective wolf' (McDowell, 1995, p. 153). Following this, reason is described as that which:

> … enables and even obliges us to step back from it [our nature] (McDowell, 1995, p. 154)

There is here the notion of both reason and reflection and they have been confused. They are both concerned with thought but this is not to say that there is no distinction between them. Reflection cannot take place without reason but reasoning can take place without reflection. The crux of the distinction is that in reflection we step back from our thought and review and evaluate. That is, reflection involves thinking about thinking or, as the behavioural sciences would have it, meta-cognition. However, I am wary of accepting meta-cognition as a synonym for reflection. Meta-cognition is a word that operates within a particular type of inquiry and can be expected to have meanings associated with that inquiry which are not associated with the ordinary language use of reflection. Meta-cognition has occupied the attention of those in the field of educational psychology in recent years. I shall confine myself to ordinary language usage.

The distinction between reasoning and reflecting is evident in the English language. We may say that we dispute, disagree with or reject a line of reasoning, but it sounds odd to say the same of reflecting. We may say that we should reflect on something but to say the same of reasoning is odd. We reflect 'on' but reason 'about'. We 'check' reasoning but not reflection. Reasoning can be valid or invalid but not reflection, though it might be irrational. None of this divorces reasoning from reflection.

With reflection, in the first instance, we return to previous thinking or thoughts; we revisit them. This is a necessary precursor to giving attention to that past and reflecting on it. If I reflect on the past I am recalling incidents that may or may not involve myself. I may make a judgement that a wrong move was made or a wrong direction taken. On the other hand I may pinpoint a decision in the past with which I am pleased because it produced a successful outcome. I may, however, merely recall the past and, as it were, run it before my mind without making judgements but simply noting that it had occurred and observing the links in the chain of a personal narrative. This is recall which does not necessarily involve reflection. Is it possible to reflect on the future? The immediate response is surely not, for the very root of the word 'reflection' is connected with the past. It may be said that I may reflect that certain possibilities may or may not arise in the future. I may reflect that this decision now will close or open a door in the future. But reflect on the future?

On this account reflection is the giving of attention towards past thinking. To reflect is a transitive verb. There must be something to reflect upon, we do not simply reflect. And this can be further broken down into the giving of attention to:
- thinking and this may be in the distant past or that just gone;
- the content of that thinking; and
- the evaluation of that thinking.

In all of these the reflection may be concerned with reasons and reasoning.

But to confine the analysis of reflection to the past is perhaps a mistake. What is the point of reflection? We step back and consider and concentrate on the past in order to do a number of possible things, among them:
- to correct previous thinking or plans; and
- to revisit our plans so that we may plan further.

The first presupposes that something has gone awry and we need to rethink the present and the possible future. The second presupposes that we have completed or nearly completed previous plans and need to consider the next move. The introduction of the notion of planning may seem at odds with the assertion that reflection is concerned with the past. Any planning incorporates a future dimension. It also incorporates the ability to imagine the future. Reflection about past or current plans necessarily incorporates a time factor.

The first step in planning may be an attempt to establish the possibilities and evaluate them. For instance, when confronted with a situation the like of which we have not met before we typically step back from the situation and make a number of possible moves. We may try to determine whether or not masterly inactivity is called for or if we must act. Both masterly inactivity, as the phrase implies, and action, demand an assessment of the results – what happens if I do nothing? What happens if I do this? We speculate. We are in the position of McDowell's reflective wolf. If we decide to do something then foresight and imagination are involved and planning. But foresight and imagination are also implied in masterly inactivity. We may attempt to find a suitable comparison between something from our past or the past of others for the situation which confronts us. This is a situation akin to finding an analogous theory in science to use in tackling a new problem. This is a reflective activity. Reflection is necessarily connected with the past and may be connected with the future. Planning necessarily involves reflection and is thus involved with both past and future. The element which takes reflection into the future is imagination. To give attention to the future we must imagine it. Once we have imagined it then it is possible to reflect upon it. Though in saying this there is the appearance of a step by step process and it is a mistake to think it so. Reflection is both retrospective and prospective.

In order to reflect we must concentrate, that is, we must single out and consider certain aspects of what we have thought or planned or done and we do this by focussing on those aspects to the exclusion of others which may be extraneous and irrelevant. Faced with what ought to be done we have to concentrate on the problem if we are to solve it. Simple attention will not do.

Consider some examples in order to unpack further the concept of reflection.

The years 1845-1849 were the years of what the Irish call 'The Great Hunger' and the English call 'the Irish Famine'. The population of Ireland in 1841 was thought to be 8,175,124. The expectation for 1851 was 9,018,799 given the normal rate of increase at the time. The census of 1851 placed the population at 6,552,385. It is estimated that 1,500,000 emigrated to England, North America, Australia and New Zealand between 1845-49. This leaves a shortfall of approximately 1,000,000 in the anticipated figure for 1851. The million are the deaths from the famine. The deaths were, in the main, due to starvation and typhus caused by the conditions of the famine (Kee, 1982).

The Great Hunger was caused by the failure of the potato crop upon which the Irish poor depended for their existence. They existed by hiring out their labour in return for a small plot on which they could grow potatoes. In many respects, for the poor, it was a cashless economy. The failure was caused by *Phytophthora Infestans*, a fungal disease which struck every year from 1845 to 1849. The disease did not effect other crops and, in fact, the years 1845-49 saw a rich harvest of grain and excellent conditions for farming livestock. The Irish poor lacked the money to have access to this food. Even in normal times a quarter of the Irish population, approximately 2,000,000, lived between potato harvests in conditions of semi-starvation.

The Irish Parliament having been abolished by England in 1800, Ireland was ruled by the English government in London. The political ideology at the time of the Irish Famine was known as 'political economy'.

> Certain almost unshakeable, sincerely held economic beliefs were to underlie all governmental policy towards the Famine. And the greatest of these was that principle of political economy which maintained that you should interfere to the absolute minimum with the market forces of supply and demand because if you *did* so interfere, you endangered the natural flow by which supplies could reach the market. (Kee, 1982, p. 82)

Implementing this principle resulted in a continued large export of food from Ireland to the continent of Europe during the famine years and the control of any relief for fear of affecting the market.

The causes of the Great Hunger and the deaths of so many are clearly a complex and difficult subject to explore. If, for the sake of argument, we cease exploration at this point and consider just the above, then, what might be said?

There are causal links between the deaths and a number of factors. These links can be analysed biologically, politically, culturally and economically. Reasons can be given for the links. What happens if we stand back from the causal network that has been constructed and evaluate it? It might be said that one sense of reflection occurs when we stand back and examine the validity of the causal links. That is, we look at the evidence for causation and evaluate that evidence. We might, as a result, give weight to some causes over others. We might weigh the strength and weakness of the overall case and this might point the way to further avenues of investigation. However, it seems to me that it is a mistake to call this reflection. It is more properly called reasoning.

Reflection is the thinking in which reasoning is involved but which goes beyond reasoning. We can stand back and evaluate the acts and omissions of the actors in the historical drama. For instance, we might say that the ideology of 'political economy' was morally bankrupt and for the English government to act on the principles of 'political economy' in the way that they did was immoral. We might contrast the government's actions with the Christian virtues of charity and benevolence towards the poor and their needs, which the English government publicly professed. In this we are engaged in making judgements based on criteria which are embedded in our ontological and epistemological beliefs. This I would consider merits the term 'reflection'. To reflect is to ponder, to mull over. It is to go beyond the construction of an argument and the checking of that argument. It is to step back and evaluate, and, because the Great Hunger was concerned with what happened to and was done to human beings, the evaluation is fundamentally a moral evaluation. In making such an evaluation we measure what happened against what we think ought to have happened. This is to place the matter before criteria which are fundamental to our conception of the world in which we exist. Reflection goes beyond but includes reasoning.

Consider a different example. Crystals take regular forms. They have a certain symmetry. Fluorite is an octahedron. The snowflake crystal has a six-fold symmetry. Iceland spar is rhomboid. In chemistry the student may learn to recognise the various shapes and link them to their chemical composition. Crystals may be grown in classroom experiments. A wealth of detail may be acquired, arranged, analysed and catalogued. It is possible to view the analysis and the catalogue arrangements to see whether or not they have any validity. However to reflect on the appearance of crystals is to confront questions about reality. Why is it that the crystals have flat planes; why is that the fluorite crystal is an octahedron? Pursuing these questions may lead to the thought that the atoms of a crystal are governed by the properties of three dimensional space. Repeated symmetry is only possible is certain ways. Crystals are natural kinds and the atoms of the crystal are examples of fundamental natural forces. Reflection leads us to considerations about the nature of the world in which we exist.

The production of the spectrum from white light has been known for millennia. For some time it was thought that the spectrum was produced by the modification of white light. White light which passed through the thinner part of a prism was modified less than that which passed through the thicker part. If we reflect on that explanation then it would appear that another prism should produce further modification. It does not. Another explanation was needed. Newton provided it. The white light is separated into constituent parts, not modified. The reflection lies in asking why it should be that white light produces a spectrum under certain conditions. Science is the attempt to provide a description and explanation of the underlying structure of nature. This is a description of a reflective activity and reflection in science takes place whenever we ask questions that bear upon the underlying structure of nature; whenever we ask the ontological questions.

The Greeks produced buildings in which columns were connected by and supported a straight beam. Their buildings are rectilinear. The Romans invented the arch. Their arches were semicircular. The physical advantages of the Roman arch over the Greek beam can be explained by some elementary physics. The results of the advantage can be seen all over Europe. At some point around 1200 A.D. an oval arch appeared in Europe. This had advantages over the Roman arch which again could be calculated. The advantages resulted in Gothic architecture. It is not known how humanity came to progress from one to the other. Was the idea sparked by the humble hen's egg which is Roman at one end and Gothic at the other and whose strength lies on the plane between the ends? This is to reflect on the historical narrative and to speculate as to its causes.

Students may be asked to solve problems in physics which involve calculations - inclined plane problems for example. To solve the problem they will have to recall how to analyse the forces involved and the formulae to apply. Having completed the calculations, being good students, they will check their work. None of this involves reflection. Reflection, in my view, enters the arena when students place the problem within its history. Reason and reflection are connected. Certainly we can and do reason and reflect on particular issues but I would argue that the appearance of reflection signals the appearance of something wider and more important than the issue of the solution of this particular problem that confronts us here and now and the checking of that solution. Reflection always brings with it the activity of placing the thinking in the context of the narrative and the context of the nature of reality. This is the case in science.

In order to reflect we must place the thought or the plan or the deed in context. All thoughts, plans and deeds have a history, a narrative which leads to their existence and without which we cannot understand fully the thought, plan or deed. This we need to bring into focus, to concentrate on, if we are to reflect with the possibility of success. Consider $T = 2\pi\sqrt{l/g}$. There is no need to itemise the meanings of the symbols to some people. They recognise it as the equation describing the motion of the simple pendulum. Some of those, a smaller number than the original set, will recognise the equation as arising within a limited system.

That is, they will recognise that there are specific limits set with regard to the arc of swing, friction, elasticity and so on. A smaller number still will recognise it as part of a revolution in physical science. At what point might it be said that a person reflects on the equation? More than merely recalling the connection of the equation with the simple pendulum is required. More than recalling that if the equation is to be derived from a simple pendulum then limits must be imposed on the system. This is simply a matter of mathematical practice. I would suggest that reflection is the proper description for that which takes place when, with recall as a base, the equation and the physical phenomenon start to be seen in the revolutionary context of which they form a coherent part. It is at this point that we talk of 'mulling over', 'pondering', 'appreciating'. This is not to suggest that recall is always a necessary base for reflection.

Plans and deeds are necessarily teleological. They are formed or performed with a purpose. Thoughts are too, though the connection is not so clear. However, the fact that we have in the language the phrase 'idle thoughts' seems to indicate that the norm is thought which is not idle but formed towards a purpose. The above points all find their expression in the language.

Reflection is essentially a teleological enterprise and a very human enterprise. It is this capacity which, in my view, is a far better possibility as that which distinguishes us from the other animals. On reflection, reason will not do!

Earlier I had argued that, following MacIntyre (1999), the capacity which may be distinctively human is expressed in at least four things:

- the ability to distance ourselves from our beliefs, decisions and actions both in the past and the present;
- the ability to evaluate our beliefs, decisions and actions;
- the ability to imagine and attempt to choose, which presupposes evaluation, our future; and
- the ability to imagine, which presupposes evaluation, our past.

The four things above express the capacity to reflect. The word 'speculate' illustrates these elements. The Macquarie Dictionary says that to speculate is:

> 1. to engage in thought or reflection, or meditate (oft. fol. by *on, upon* or a clause). 2. to indulge in conjectural thought. (Deldridge, 1987)

Speculation in the above sense often begins with the query – what if? What if the earth moved round the sun; what if I were to ride on a beam of light looking back at a clock; what if the continents are floating and can move? Not all 'what if' questions lead to speculation. What if I switch off the electricity connection to this computer? It will shut down. The speculative queries are those which lead a person to question the basis on which he or she views reality. The speculative questions demand the exercise of the imagination. In this sense it becomes clear that speculation is the life blood of science. Indeed, it becomes clear that speculation is the life blood of all attempts to advance our knowledge.

The concept of 'regret' provides further illustration of reflection. We may regret the past, the present or the future. In regretting we always admit that things should have been and should be managed differently while at the same time admitting that this might not be possible. There is always evaluation and that evaluation is against what we conceive as the way the world ought to be. In regretting the past we recall what has happened, we imagine what might have been different and we evaluate what has occurred. In regretting the present we see how it has come to pass and how it may proceed. In regretting the future there is an element of helplessness in the face of what we see impending. To regret is a reflective process which requires reasoning, imagination and evaluation.

Reflection is a capacity which cannot be exercised in a vacuum. A human child goes:

> … beyond the reasoning characteristic of dolphins when they become able to reflect on and to pass judgment on the reasons by which they have hitherto been guided. This transition is one that dolphins have not made, so far as we know, but we can learn a good deal from them – and from chimpanzees and members of various other species – about the preconditions for making it. (MacIntyre, 1999, p. 57)

The acquisition of the language of reflection is a pre-condition of rationality (MacIntyre, 1999, p. 54). Rationality is only intelligible in the context of the ability to reflect and consider whether or not this or that reason is appropriate or relevant. And this rationality would not be possible without the language that enabled it. Rationality and reflection are intimately associated.

I have argued that the hard core of beliefs of a culture define rationality for that culture and thus for the individuals within and part of that culture. The principle of political economy held by the English Government in 1846-50 can be classed as part of the hard core of beliefs against which proposed actions were measured. The principle defined what was to count as reasonable, evidence and sound judgement in a particular area. The principle is an ontological and epistemological dictum.

The hard core of beliefs of an individual define for that individual what are to count as reasons, evidence and sound argument. They may also be said to start to define the individual. The flat-earther looks at the world in one way, the spheroid earther another. Jane Austen looks at the world very differently from Graham Greene. When Austen and Greene engage in reflection they do so against the way in which they see the world. Their books are a reflection of and on that world. The reflection takes place in the context of the hard core. Rationality is only intelligible in the context of the ability to reflect and rationality is defined by the hard core. The hard core is defended from attack by the erection of auxiliary hypotheses which counter the opposing arguments.

Consequently, we may say that there is a capacity to reflect which, for its exercise and development, requires a hard core of belief which is the basis of the ontological and epistemological beliefs of an individual. As young animals we have certain basic desires and needs. We have a variety of instincts, though perhaps not many compared with other animals, which help us to satisfy those needs. We also appear to have been hard-wired in such a way that we are able to discriminate the world of our experience on the basis of natural kinds (Markman and Gelman, 1986; 1987; Kornblith, 1993). From this beginning we develop the ability to find a reason for action in the satisfaction of our needs and desires and it seems evident that we need something like this beginning in order to develop so. We can do this rather than that because this is more productive of the satisfaction of our desires. On an evolutionary basis it would appear that we are performing as the higher animals perform. We start to differ, as far as we know, from the animals when we start to develop the capacity to stand back from our needs and desires and judge whether or not they are worthwhile. We reflect rather than simply reason. The initial notion of worthwhileness is with reference to what it is worthwhile for me to do, what it is best for me to desire. And this is assessed against my small world and my place and the place of others in that small world. To put it another way, assessed against a developing hard core. I take account of others not merely myself. With development this world enlarges and the measure of what counts as worthwhile and rational enlarges and becomes more complex encompassing reflection on the past and consideration of the future. This is a process of education. The development of a reflective human being is not the development of, on the one hand, a capacity, and, on the other hand, a hard core of ontological and epistemological beliefs. They are interdependent. The aim of both informal and formal education should be to develop both. The capacity to reflect is developed through encouraging the child to imagine, to speculate, to ponder, to evaluate. The hard core is developed through the formal and informal processes by which children acquire their view of the world and their place in it. It is a mistake to think that the hard core is composed of empirical beliefs and moral principles. The human being is not that simple. Religion, myth and beliefs which are fervently held but for which there is no evidence may all be present. The 'Dreaming' of Australian Aboriginals plays a very real part in the hard core of those peoples. The Icelandic sagas still form part of the hard core of that culture (Smiley, 2000). The question for the beginning of 2005 by the publisher of the scientific website *web.org* is illustrative of fervently held beliefs for which there is no evidence:

> What do you believe is true even though you cannot prove it? (Brockman, 2005)

And among the published responses of 120 scientists and thinkers is that from the physicist, Paul Davies, who believes that the universe holds other life forms though he cannot prove it:

> … I believe we are not alone because life seems to be a fundamental, and not merely an incidental, property of nature. It is built into the great cosmic scheme at the deepest level, and therefore likely to be pervasive. I make this sweeping claim because life has produced mind, and through mind, beings who do not merely observe the universe, but have come to understand it through science, mathematics and reasoning. This is hardly an insignificant embellishment on the cosmic drama, but a stunning and unexpected bonus. Somehow life is able to link up with the basic workings of the cosmos, resonating with the hidden mathematical order that makes it tick. And that's a quirk too far for me. (Brockman, 2005)

And in that reply it is possible to see elements of Davies' hard core.

Wertheim has argued that physics has always had a connection with religion. Discussing Einstein's view of the connection between physics and God, she says:

> One of his [Einstein's] most famous aphorisms is the oft-quoted "God does not play dice," which pithily summed up his objections to quantum mechanics. But there is another English translation of the same remark (originally uttered in German) that captures more fully the thrust of Einstein's thinking – "God casts the die, not the dice." Here was Einstein's philosophy of nature in a nutshell: The universe is divine creation, and it is the task of the physicist to discover the mathematical die from which it was cast. It is this essentially Pythagorean dream that inspires Stephen Hawking and others today. (Wertheim, 1997, pp. 185-186)

The hard core will be subject to attack either from others or from an individual's own observations of reality. The response to attack is the defence of auxiliary hypotheses. This is not to say that there can be no change in the hard core of a culture or a person. Clearly this can happen and does. History is witness to that.

The narrative of a human life is called upon when we reflect on whether or not this should have been done or believed or whether or not we should do this or that, believe this or that. Reflection here involves who the human being is and his or her stance to the world. Reflection in any practice, whether it be science, fishing or painting brings into play, for the reflecting human being, that human being's connection with and view of, the world of his or her existence. The examples of reflection above indicate that the attempt is to make and retain a coherent, consistent account of the world of our existence. It is to seek harmony.

This consideration of reflection began with an investigation of the concept of human flourishing. It is now possible to say that what a human being needs to flourish is to develop that capacity by which human beings are distinguished from the rest of the animal world, the capacity to reflect. This capacity must be the prime objective of education both formal and informal.

CHAPTER 16

THE HARD CORE

I have argued the connection between the hard core and rationality and between the hard core and science. There are other issues concerning the hard core. Human beings are social animals who have the distinctive capacity to reflect. The development of this capacity is central to their flourishing. Because they are able to reflect, as the capacity to reflect develops, the individual necessarily asks questions about the world of his or her existence. The non-reflecting individual does not see himself or herself in the context of the world. They do not flourish. An individual's questions about the world and his or her place in it may and should be encouraged by society. Education, both formal and informal, has an important role to play. For instance, the student may be confronted with:

Tomorrow; and tomorrow, and tomorrow,

Creeps in this petty pace from day to day,

To the last syllable of recorded time;

And all our yesterdays have lighted fools

The way to dusty death. Out, out brief candle!

Life's but a walking shadow; a poor player,

That struts and frets his hour upon the stage,

And then is heard no more: it is a tale

Told by an idiot, full of sound and fury,

Signifying nothing. (MacBeth Act V Sc V)

MacBeth's reflection on the world as he sees it goes to the core of his being. It is this core that has to be developed if the individual is flourish. Whether or not MacBeth's view of reality is tenable is a question for the reflective human being and this is to admit the possibility of criticising a person's hard core. We clearly do. All subjects taught in schools have within them the questions that arise from reflection and can only be answered by reflection. Some subjects such as history and science are clearly reflective subjects, but, in fact, all those subjects which we have traditionally called disciplines are reflective subjects. However when we teach about the world and our place in it whether through academic disciplines, integrated areas or crafts and technical areas, the reflective questions are there not merely to be used but to enhance the area of study and aid the individual to flourish.

The types of questions that a developing reflective human being may ask are driven by the context of his or her existence. That context is a context which is both physical and social. That is the individual exists in a natural world which has within it a social world. These questions would include:

- What am I doing here?
- What is my relationship to others?
- What is my relationship to the physical world?
- What is the nature of the physical world?
- What is the nature of reality?

These questions lead to others. The first to spiritual questions; the second to questions about dependence upon and competition with others; the third to ecological questions; the latter two to particular ontological and epistemological considerations. Further questions again arise from each. I grow up dependent upon others and later I become independent. Or is it better to say that dependence becomes inter-dependence? The world in which I live believes in competitive individualism. Is it possible to reconcile this with interdependence? What is the right thing to do or believe?

In attempting to frame and to answer these questions the individual must reflect. They are questions which, as far as we know, do not occur to other animals. They are questions whose answers come to constitute the hard core of the individual. We talk of the core of a person and by this we refer to his or her central beliefs, to his or her fundamental view of the world. The use of 'central', 'fundamental', 'core' and other similar words are used to indicate that these beliefs are constitutive of the identity of that person. They define him or her.

To an extent the hard core is contextually determined. The question whether or not my relationship with others should be on the basis of competitive individualism arises in a social context which admits of or knows of this possibility. There has been a significant amount of research into living generations in Western society. The Baby Boomers is the generation born between 1946 and 1964. They were followed by Generation X, 1965-1979, Generation Y, 1980-1994, and Generation Z, 1995-2009. In comparing these generations it is important to neither ignore differences nor minimise them (McCrindle, 2006).

McCrindle points out that the economic, social and political conditions in which each generation lives differ and the differences serve to divide and distinguish (McCrindle, 2006). The economic, social and political conditions that a generation experiences influence the way in which they see the world and the beliefs that they come to have about the world and their place in it. That said, it is important to note that the change from one generation to the next may be minimal or it may be large. In periods of economic, social and political stability there would be an expectation of little change. Instability would foster change. McCrindle quotes an ancient saying:

> People resemble their times more than they resemble their parents. (McCrindle, 2006)

This may be true with the caveat that there may be little or great differences. These ideas are not something new. The narrative of history enables us to see that former generations are different in the way they organised society and the beliefs that they had about the world in which they existed. Hard cores change. As we gain knowledge and better explanations of the world around us so we are better placed to have a justifiable hard core. The affair is a quest.

While the ideas of change are not new it is sometimes difficult for an older generation to realise the different context and its effect on the succeeding generation. This is particularly a problem for parents and teachers. For instance, in Western society, today's adolescents are faced with different challenges and have different priorities. Celek and Zander argue that they are more concerned with developing relationships than achievement (Celek & Zander, 1996). They appear to have less connection with traditional religion and more with various forms of spirituality. All of this is an indication that the world view of the current Western adolescent generation is distinguishable from the world view of their parents. We may generalise this beyond the current Western adolescent generation and say that for parents and teachers there is a need to understand the changed priorities and questions of the generation they are raising. Only when this is achieved can they embark successfully on development of the capacity to reflect and the achievement of human flourishing. This, however, has to be combined with an appreciation of the human narrative and human achievement. If the past is lost so is the present.

If generations are to flourish as human beings then they must reflect about the physical and social world in which they live. That concern must be a quest for the best possible view at their time viewed in the context of the narrative of human history. This might seem to indicate that the matter descends into the relativistic state against which I have argued. This is not so.

I have argued that the touchstone is always reality and this is independent of our thoughts about it. I shall argue that the other things that do not change are the necessity to develop the capacity to reflect and the cardinal virtues which are necessary for that reflection.

CHAPTER 17

THE VIRTUES

I have argued the connection between the concept of reflection and the distinctive human capacities that MacIntyre (1999), picks out:

- the ability to distance ourselves from our beliefs, decisions and actions both in the past and the present;
- the ability to evaluate our beliefs, decisions and actions;
- the ability to imagine and attempt to choose, which presupposes evaluation, our future; and
- the ability to imagine, which presupposes evaluation, our past.

Consider each item in turn.

When we distance ourselves from our beliefs, decisions and actions we stand back, we attempt to view them from a more disinterested perspective, a more neutral position. Why? The assumption is that if we can disengage ourselves at least to some extent from situation then there is a better chance of assessing a belief, making a decision, or acting that is not unduly swayed by considerations which are irrelevant to the belief, decision or action. The disengagement is a necessary prelude to any successful evaluation. However, the evaluation only becomes a success when we have the courage of our convictions.

Distancing ourselves from our beliefs, decisions and actions is not confined to the present and immediate beliefs or decisions or actions. We have we hope, a future. This future is something to be considered, planned for, anticipated perhaps joyfully perhaps ruefully but certainly carefully. The person who proceeds through life without a thought for the future is described in terms which indicate that he or she is unusual not to say impractical and silly. We expect people to have a care for the future, to exhibit some foresight and to plan. All this involves an attempt to choose the road we shall travel. The choice cannot be made without imagining what the future might be like on this account or that, without imagining where this choice might lead us or that. Such imaginings can be humdrum or highly creative.

The imagination and choice involved in considering the future involves the narrative of our lives and thus, necessarily, involves us in the past. We are what we have been. But what we have been is a picture that is not identical with our past for our past is as we imagine it and that imagination is partial and, perhaps, skewed.

The ability to reflect and the act of reflecting involves all of this, though not necessarily all at the same time. What are we saying then that we do when we exhibit MacIntyre's four items? We exhibit a care make the best possible decision or act in the best possible way; we attempt to make judgements in a cool manner, we try and use foresight and we call on the past. There are words for these things which have been handed down to us over the centuries – justice, prudence, temperance and fortitude. It is not surprising that these old notions should still have purchase, human beings have not changed a great deal. And the old words are not to be discarded or dismissed lightly. Justice and prudence are still used though with a tendency to exhibit somewhat different meanings from previous times. Temperance became confined to drink and fortitude has fallen into disuse. To discard without the attempt to rehabilitate is discard the wisdom of the past in the centuries of thought and writing about them. As it happens the cardinal virtues have been making a comeback in the last decade. Prudence is no longer thought of as merely caution; temperance is not confined to drink; fortitude is reconnected with moral courage. The signs of the change are being registered in the dictionaries. The Oxford English Dictionary states that temperance is 'abstinence from drink'. The dictionary from a newer land, the Macquarie Dictionary, states the older meaning that temperance is 'moderation or self-restraint in action, statement, etc; self-control' as well as noting the abstinence from drink' meaning. A recent best-selling book on the virtues includes them all in their original guise (Comte-Sponville. 2003).

The old words name virtues. What is a virtue? The function of a knife is to cut and I have argued that a knife cannot be defined independently of a good knife. However, we may assert something in virtue of which this or that is a knife. Virtue and function are linked. A knife is of value in virtue of its ability to cut. A sheepdog is of value in virtue of its ability to work with sheep. To talk of the virtue of a thing or being is to talk of its distinctive worth and for the thing or being to have that virtue is to have a distinctive excellence. In what does the distinctive excellence of a human reside? I have argued that the distinctive excellence of a human being resides in he or she flourishing in a particular way and the link between virtue and reflection began to appear.

The four cardinal virtues of antiquity and the Middle Ages were prudence, justice, temperance and fortitude. Traditionally they have been considered to be the necessary possession of a flourishing human being. Consider prudence.

> Prudence presupposes uncertainty, risk, chance, and the unknown…Prudence is not a science; rather, it replaces science where science is lacking. One deliberates only when one has choice to make, in other words, when no proof is possible or adequate… (Comte-Sponville, 2003, p. 32)

Someone who has become prudent is no longer rash in making judgements or taking action for she or he has reflected on the possible consequences. If she or he has become prudent, rather than merely practising prudence on this occasion, they will find it undesirable to be rash or impulsive. It will jar with them and make them uncomfortable. It may not make them popular with those who make a living by encouraging impulses. Nor may it make them popular with those who seek to manipulate. In the English language prudence is associated with being discreet and circumspect, being careful and cautious. It is associated with the future for to be prudent is to take care or caution in what may happen in the future as the result of what is said or done. It is to be provident, that is, to provide for the future, to have foresight and not merely in relation to self-interest. The corresponding vice is that of rashness or impulsiveness, actions and thought characterised by a lack of foresight.

This description of prudence is the description of a character trait that has to be acquired if a person is to be able to reflect and come to the best possible decision. It is a partial description of what it is to be a reflective person. Prudence, therefore, may be said to be constitutive of the reflective person.

Rationality is only intelligible in the context of the ability to reflect and consider whether or not this reason or that is appropriate, is relevant. Aquinas argued:

> For the formal principle of the virtue of which we speak now is good as defined by reason; which good can be considered in two ways. First, as existing in the very act of reason: and thus we have one principle virtue *Prudence* … (Aquinas, 1948)

Therefore, we have a connection between prudence, rationality and reflection. Prudence is an intellectual virtue which is essential to the practice of the moral virtues. Prudence may be said to be the essence of reflection. It is the virtue which is necessary for any other virtue to operate. As there is this connection between prudence and rationality, then, it will be the case that what counts as prudent will be relative to the hard core of the individual or the culture. However, as in all previous similar cases, the touchstone for what is prudent is reality. What counts as prudent will change, and can be seen to have changed throughout history. It changes as the hard core is modified and changed in the light of advancing knowledge and theories and beliefs that are judged superior to those that they replace. The judgement of what counts as an advance and what counts as superior is based on reality.

To complete the above quotation from Aquinas:

> Secondly, according as the reason puts its order into something else; either into operations, and then we have *Justice*; or into passions, and then we need two virtues. For the need of putting the order of reason into the passions is due to their thwarting reason: and this occurs in two ways. First, by the passions inciting to something against reason; and then the passions need a curb, which we call *Temperance*. Secondly, by the passions withdrawing us from following the dictate of reason, e.g. through fear of danger or toil: and then man needs to be strengthened for that which reason dictates, lest he turn back; and to this end there is *Fortitude*. (Aquinas, 1948)

Consider justice. Justice brings out another aspect of some of the virtues. To act justly is to give to each according to their need and to expect from each according to their ability. We may recognise the needs of others but ignore them. We may recognise our own ability but not use it in the service of others. Typically the unjust person is one who is driven by self-interest. In pursuit of that self-interest we may lie, cheat or steal the reputation of another. The human infant is typically driven by self-interest and it is only as that infant develops the capacity to reflect that he or she may come to realise the needs and abilities of others and the dependence that he or she has on them. The human being then starts to acquire the virtue of justice and this is a virtue which is constitutive of the reflective person.

Consider temperance. Temperance brings out another aspect of some of the virtues, that they arise to give human beings some control over their nature. Human beings like all animals are subject to desires and wants. In the infant these desires and wants are immediate causes of action. The infant seeks to satisfy them. But we recognise that not all desires and wants are worthwhile. There are some desires that we should satisfy and others that we should not. Temperance is the virtue that enables control over illegitimate desires and promotes those which are worthwhile. The intemperate person is one who attempts to satisfy all his or her desires without thought for the consequences to either themselves or others. So, once again, temperance can be seen as constitutive of the reflective person.

Consider fortitude. Fortitude is a word that, like temperance and prudence, has fallen into disuse. As a result their meanings have been forgotten to some extent. It might be thought that courage is a synonym for fortitude. Certainly courage is universally admired and, its opposite, cowardice despised. But fortitude is a virtue whereas courage is not. Courage may be the possession of a thief or a murderer. It is morally neutral and takes its moral attribution from the context in which it is displayed. Comte-Sponville (2003, p.49) argues that fortitude is courage in the context of altruism and that its core is moral strength. All virtuous action requires effort, sometimes more, sometimes less, to overcome the human inclinations that we are heir too. Fortitude as moral strength, as moral courage, as firmness of intent, is a virtue which is necessary to the exercise of all the virtues. Comte-Sponville follows Aquinas in this when he comments:

> Thomas Aquinas explains this very well: *fortitudo* (moral strength, courage), like prudence but in a different way, is "a condition of every virtue" and also, in the face of danger, simply one virtue among many. A general as well as a strictly cardinal virtue, it supports the others like a pivot or hinge (*cardo*). (Comte-Sponville, 2003, p. 50)

Fortitude, on this account, is an essential element of the reflective person.

These and other virtues are derived from what human nature is like. The virtues must be acquired if we are to fulfil our capacities as human beings not merely in the singular sense but as a social being. Consequently, they are constitutive of the good human being. They are constitutive of human flourishing. They are also constitutive of ethical action. This is clearly an Aristotelian view of the matter though it may also be said to be, in some respects, a Confucian view of the matter. Aristotle starts with an unrealised human being and the possession of the virtues is the bridge by which a human being realises his *telos*, that is, his or her capacities as a flourishing human being. We learn them from our family first, then society and it is an interesting question as to what form of formal education is best suited to the development of the virtues. Certainly it is not clear how or whether some part of our education should be set aside for specific education in the virtues. It is clear that some part of formal education should be set aside for the development of the hard core.

It is important to note that the virtues are not dispositions but character traits. A disposition is a tendency to act in a certain way and to possess a virtue is much more than this. It is not a tendency for tendency admits of the possibility of acting in a way that is contrary to the tendency. The possession of virtue does not admit of this. We have a phrase that indicates this. Of an honest person who avoids the dishonest act we say – it never crossed his or her mind. There are two important points here. The 'never' indicates that the possession of the virtue of honesty admits of no possibility of dishonesty. The 'mind' indicates the reflective, personal nature of the virtue which is part of what constitutes this particular person. A character trait is part of what constitutes a particular person and the sum of the character traits of a person are that person. When we say of a person that they are honest, or just or prudent then we are saying that they possess those virtues and that they live in a certain way. To characterise a person as honest, just or prudent is to say something about what constitutes them, who they are. On this view ethics is no longer a matter of personal preference but a matter of objective necessity for human flourishing. It is not a matter of personal construction.

This analysis of the virtues and their relationship to human nature and capacity now links with the initiation of a human being into a practice such as science. Science is necessarily a reflective activity. It also links with the criticism of constructivism for if constructivism has a problem with reflection then it has a problem with teaching and learning a practice such as science, a problem with the acquisition of the virtues and a problem with a child's growth into a good human being

CHAPTER 18

SCIENCE, THE VIRTUES AND THE HARD CORE

I have argued that the criteria internal to the practice of science are those criteria, and they include values, which are the means by which the worth of scientific work is assessed as scientific work. There is no entailment that leads to a rejection of the applicability of social, political and religious values. Nor is there any necessary division between the set of scientific values and those external to the practice of science.

Scientists practise science. They practise physics, chemistry, biology, astronomy and many more. I have adopted MacIntyre's analysis of practice, narrative and virtue, though any misinterpretation is my own (MacIntyre, 1981, 1988, 1990a, 1991, 1994).

Chemistry is a complex and coherent activity. It is a socially established and cooperative social activity conducted by human beings. Chemistry can be pursued for fame, power and profit. These ends are external to the practice of chemistry and, if chemistry is pursued solely for those ends, then, the practice of chemistry is controlled by those ends. So the ends of power and profit may on occasion dictate that the chemist keeps his or her work secret. The end of fame may lead to a distortion of results. The end of profit may lead to a disregard of the end use of the practice. The point here is that external ends have no investment in the practice but only in themselves and thus the practice becomes merely a tool to be used. The external ends are not specifically related to any practice but to all practices. This, of course, is not to say that fame, power and profit may not on occasion be worthwhile ends, it is merely to say that they are external to the practice of chemistry. Such external ends are typically ends that are to be sought in competition with others and the end result of that competition is that whoever acquires fame or power or profit in doing so lessens that which others can acquire. Fame, power and profit are possessions of the individual or the group which are held in competition against others and other groups.

What are the goods internal to the practice of chemistry? There is first of all the product of the practice. When Mendeleev constructed the Periodic Table he produced a product which was excellent. How was it known to be excellent? Because it confronted and solved a problem which faced chemists at the time of Mendeleev. Problems do not arise *in vacuo*. That problem occurred at a certain point in the history of chemistry and problems do not arise and cannot be understood as problems except in the context of their history. The elements with which Mendeleev worked were those that were known at the time and they had come to be known over time. Eventually there were sufficient elements known for Mendeleev to seek a pattern. Mendeleev produced a solution that at one and the same time solved the problem and showed the future direction of research. The excellence was determined by reference to the history of chemical knowledge and theory not by reference to external factors. That the solution also brought Mendeleev scientific fame was a matter external to the practice of chemistry. That the solution enabled various commercial advances to be made was external to the practice of chemistry.

Galileo practised physics, among other things. His mechanics departed from that of Aristotle. In doing so it encompassed Aristotelian mechanics, removed flaws, added new ideas, explained more. By the standards internal to physics the work of Galileo on mechanics was excellent. The history of Galileo's fame and fortune as a result of his work in that and other areas is not one of unalloyed success. At times he was a commercial success at other times a religious outcast. The vicissitudes of his life were external to the excellence of his physics.

Miller's (1994) error is to consider the pursuit of scientific truth as external to the practice of a science such as physics. Frankel (1987) has written a succinct history of the continental drift debate which is pertinent. In the early years of the twentieth century the majority of earth scientists held the view that the continents of the world had suffered no horizontal displacement during their existence. This was the stage that the earth scientists had reached in their pursuit of knowledge about the earth. There were a number of problems with the theory. It was clear to schoolchildren that there was a fit between Africa and South America. There was evidence of geological similarities between North America and Europe. It was difficult to explain why mountain ranges tended to appear on coast lines. Perhaps most of all, it was difficult to explain the disjunctive distribution of flora and fauna throughout the earth. Various attempts were made to overcome the difficulties. Sinking of the land between continents explained the separation of continents, land bridges explained the disjunctive distribution of flora and fauna.

Wegener (1924) proposed that the continents were a light sialic layer floating on a heavy basaltic layer and moving through it and over it. The continents could, had and were drifting. Earth scientists began to divide into two camps – those who accepted a static view of the continents and those who accepted continental drift. The drifters claimed that, in the case of the flora and fauna, the land bridges were not a good explanation. On the other hand they claimed that only drift theory could account for the disjunctive distribution of flora and fauna. They pointed to the example of Australia. The oldest flora and fauna in Australia has affinity to that found in Sri Lanka and India. The next oldest has affinity to that found in South America. These were the marsupials and the affinity resulted from the time that Australia was attached by way of the Antarctic to South America. The Australian placental mammals, the youngest of the Australian fauna, had affinity to those found in Asia and arrived in Australia by way of the islands to the north of Australia.

The major problem for the drifters was the mechanism which propelled the continents. Polar-flight and tidal force were considered as possibilities. But they did not seem adequate and it did not seem possible that the continents would have survived the movement. By 1949 there was an impasse though drift theory was still a significant impetus to research. Finally in the late 1950s paleomagnetism entered the debate and when combined with marine geology finished the debate in favour of continental drift. The new theoretical approach was a significant problem for all earth scientists except the minority who had had some training in geophysics.

This brief account highlights a number of features of scientific argument and progress.

In every case, whether we think of the work of Newton or Boltzmann, Dalton or Mendeleev, Aristotle or Galileo, the work of the scientist as a practising scientist was judged against the situation they confronted, the history of the science to that point and the contribution which they made to correct flaws and thus render the science more coherent, more explicable of reality and direct the future of the practice. The evaluation was internal to the practice. An example from this century is the progress of astro-physics.

The product of the work of the practising scientist is a gift to the practice as a whole. That is, all practising scientists benefit from it. The product is an advancement of that practice and, as such, necessarily benefits all who practice. Without that gift practising scientists cannot know the history of the science to the present and thus are in no position to understand fully the current dilemmas and problems. To keep secret a discovery or a theory is to attempt a disruption of the narrative.

The product is necessarily seen against the history from which it grew and to which it is a reaction. It is judged against the problems that confront the practice at that time. Its concepts are to be understood by means of the concepts that preceded it. Therefore, it is the case that only those who are sufficiently aware of the history of the problem and the solution to it are capable of understanding the worth of the product. Practitioners alone are capable of appreciating the internal worth of the product. There is nothing in this which precludes students from becoming sufficiently aware to appreciate the worth of the science they are studying. There is nothing in this which precludes students from being initiated into the practice. There is everything in this to argue that a system, which places the students on the outside of science looking in and learning about it, is a misadventure. This capability, while it separates the non-practitioners from the practitioners, must also be seen as dividing practitioners. Not every practitioner of science is capable of recognising and appreciating the worth of the latest product in nuclear physics or archaeology. Scientists are trained as, and practice as, specialists in a particular area of scientific practice. They therefore tend to be limited in their ability to appreciate the worth of particular scientific products. But this is not to say that scientists are limited in a similar way in their ability to appreciate the process rather than the product. Nor is it to say that non-practitioners are not in a position to appreciate the process rather than the product.

What has been distinguished is the evaluation of work either by the standards internal to the practice or by standards external to the practice. Those who wish to practice chemistry necessarily have to submit themselves and their work to the standards internal to that practice.

However, in submitting to the standards internal to the practice the scientist is making a particular commitment. He or she may not produce the work of a Galileo or a Boltzmann, their work may be mundane puzzle solving. They may work in a commercial laboratory for a drug company performing tasks which will never lead to more than their pay package but, in so far as they perform their work with a commitment to the standards internal to the practice of the science then they are engaged in a form of life which is, in itself, an internal good of the practice of that science. This is what it is to be rational as a scientist. The norms to which the scientist commits himself or herself are those norms which constitute rationality.

In talking of the commitment which a practising scientist makes or the initiate is learning to make we are talking of the acquisition not simply of rules or principles on which the scientist ought to act but of something more fundamental, the acquisition of what it is to be a scientist and the development of a human being. This takes us into a difference between the ethics of virtue and rule-orientated ethics such as act-utilitarianism and straightforward deontological ethics. A virtue is not a principle of duty. 'Scientists ought to act honestly with respect to their experiments' is a principle of duty. It stands outside the person who may or may not act upon it. A virtue, on the other hand, is a quality or trait of the person which he or she may or may not possess and may or may not choose to seek to acquire. On occasion, in English, the distinction between principle and virtue is made by using different but related words. So benevolence is a virtue and beneficence is a principle. I have said that principles of duty stand outside the person and that virtues do not. This is crucial. If the sources of action as a human being or as a scientist are not to be external or adventitious then they must be connected to character traits. The scientist, acting as a scientist and thus seeking scientific excellence, must necessarily be driven by the virtues crucial to his or her practice. The same is true of the scientist as he or she exists within a community of scientists and also the wider community. Moreover, the importance of the virtues in the practice of science connects with the provisional nature of scientific theory. No scientists can ever be said to know the truth for no theory is more than provisional. Thus no scientist ever knows all the relevant facts. In such a situation it is not possible to demand that the scientist acts according to rule or principle. All that can be insisted upon is that the scientist acts out of those traits which he or she has acquired in coming to be a scientist.

MacIntyre has proposed a 'partial and tentative' definition of a virtue as:

> A virtue is an acquired human quality the possession and exercise of which tends to enable us to achieve those goods which are internal to practices and the lack of which effectively prevents us from achieving any such goods. (MacIntyre, 1981, p.,191)

MacIntyre then proceeds to argue for three particular virtues in such a way that they are seen as necessary qualities for practitioners and consequently go beyond a 'partial and tentative' definition.

A practice, any practice, but certainly the practice of science, is a social activity. Scientific practitioners are the custodians and guardians of the standards internal to the practice of science. Any person who wishes to practise science and, *a fortiori*, those who wish to be admitted to that practice, have to submit themselves not only to the internal standards but also to the judgements of their fellow practitioners. For the internal standards are carried on and practised by the practitioners and it is they who will point out failings and the occasions on which there is a falling short of the standards. Scientists work in a community of scientists. Work is published, commented on, modified, defended. Ideas are spread abroad and developed, taken further and progress made. Participation in this social activity is to participate in rational endeavours and to act rationally. Mendeleev is an example of a person from Tsarist Russia who was ignored by the rulers of his country but worked and was feted abroad within the community of scientists to which he belonged.

Two forms of criticism are important, self-criticism and that by other practitioners. Both are vital. Practitioners have to learn to accept the criticism of others and also learn to evaluate and criticise their own work. This requires the practice of certain virtues. A possible list may include - humility in having a modest sense of your own significance for, as has been said, you are standing on the shoulders of giants; fortitude in having the moral strength to criticise and to accept criticism; honesty in evaluation; prudence in the avoidance of rash judgements; justice in giving to others what is due to them. When Allchin says that scientists value honesty, and includes this as one of the values that are considered goods in both society and in science, he is talking of a virtue.

Longino talks specifically of cognitive virtues and names 'accuracy', 'coherence' and 'breadth of scope'. Hardy and Taylor talk of coupling Habermas' ideas about society and truth, freedom and justice with the radical constructivism of von Glasersfeld in order to solve the value problem. In all these cases, and I would suggest in others, the attempt to say something about scientific knowledge leads to a need for valuative decisions. Why is this? One explanation and the one to which I subscribe is that prior to the Enlightenment the Aristotelian system as it had developed in the hands of Aquinas and others had an inbuilt idea of value. Teleological explanation and the over-riding importance of human flourishing provided the *telos*. This was lost in the Enlightenment. So people attempting to say something about scientific knowledge continually come to a point at which their tradition, the Enlightenment tradition, prevents them from saying anything conclusive, anything about the value of that in which they were engaged. The result is interminable arguments. The point about these virtues is that they are qualities which are necessary to the practitioner. The lack of them prevents practice.

The lack of the virtues too widely within the society of practitioners destroys the practice. Practitioners of chemistry share the same internal standards. It is by and through such standards that chemists define their practice-relationship to one another. But the internal standards cannot be realised in practice except with the practice of certain virtues, among them humility, fortitude, prudence, justice and honesty. It is through the practice of these virtues within their practice that chemists define their relationship to one another. This is true of any practice. And, given what I have argued earlier, the practice of these virtues is connected with the practice of reflection. Reflection contains standards which are not external to the social world and the social practices within which we exist. There is not some external standard to which we may appeal to adjudicate. Reflection is part and parcel of our activities as human beings when we seek excellence.

The ancient Greeks named the virtues as wisdom, courage, temperance and justice. From a following age we have the cardinal virtues of justice, prudence, temperance and fortitude. I put to one side here the theological virtues to which Aquinas subscribed. I take wisdom to be something other than the virtues. Working from the cardinal virtues we can say that in science they must be practised. Justice includes the notion of desert and this has to be granted to and expected from those who are fellow practitioners. Scientists must do justice to the theory. Prudence is the quality of exercising cautious good judgement without failing to be courageous when necessary. Temperance is the quality of self-restraint in action and word. Fortitude is the quality of moral strength. All these qualities are those which the scientist practising his or her craft excellently must exhibit if the work that they produce is to be excellent. The one cannot be divorced from the other. The exercise of the virtues and the search for knowledge go hand in hand.

Allchin's list of typical scientific values, reliability, testability, accuracy, precision, generality, simplicity of concepts, heuristic power and novelty … (Allchin, 1999, p. 3) are those clearly related to the scientific pursuit of an explanation of the physical world. Other values he sees as social values which may appear as important in science as in the wider world. I have argued that this is a misconception. Without certain of the virtues science cannot proceed as science.

Scientists, asserts Allchin, value accuracy. But what is the attempt to be as accurate as possible but the attempt to act as honestly and as prudently as possible? The exercise of the human quality of honesty is, in relation to scientific work, whether in measurements or statements, an exercise in accurate measurement and statement. If a scientist is accurate in giving an account of a physical phenomenon, then he or she is behaving honestly. Honesty is not simply a social virtue but an intellectual virtue that demands that practitioners inspect their personal ontological and epistemological framework and see it in the context of their history in the society of which they are a part. This society is not simply the broad society in which they live but the community of practitioners of which they are a part. This task is not easy and participation requires the virtues. The possession and exercise of the virtues in science is not merely the possession and exercise of qualities between social beings in a particular practice relationship but also the possession and exercise of qualities in relation to the subject of their practice.

Allchin makes a distinction between scientific values and social values and thus, by implication, a distinction between the values that exist within science and the values that exist in the social community. In my view this is a false distinction. The virtues which are necessary for the pursuit of science are, at least in part, virtues which are necessary because of the social nature of the practice which is science. Wherever there is a social group such virtues will be called for. The virtues which have been mentioned so far are typically those which have always been thought to be relevant to the continued existence of a community. They are the qualities the possession and exercise of which are necessary for the members of the community to realise the goods internal to that community. A capacity which is claimed as a virtue within the practice of science is not a virtue unless it is also a virtue necessary for the good of the community. Virtues are not particular and peculiar to specific practices, if they were they would be idiosyncratic capacities which could be incompatible with other virtues. The virtues that are essential to the flourishing of a community clearly can give rise to problems when the community exists within a larger group whose avowed aims are external to those of the community. This arises in much the same way as it would arise where the chemist pursues chemistry for the sake of power or profit.

The thread which runs through this view of the practice of science is that of history. Modern physics, chemistry, biology, geology and astronomy are all the current states of a long historical narrative. In that narrative to understand the present the past must be understood for the present is the result of humanity's puzzlement over the past. It is impossible to understand Galileo without some knowledge of Aristotelian physics, it is impossible to understand Copernicus without some knowledge of Ptolemy, it is impossible to understand Boltzmann without some knowledge of Newton and before Newton, Galileo, and before Galileo and so on.

The narrative of the science is also a narrative of the scientist. $F = m\,dv/dt$ is a well-known formula attributed to the work of Newton. The narrative that leads to that formula is not merely a narrative of formula and argument but also a narrative of avenues considered and rejected, avenues followed to a dead end. It is also necessary to understand something of what Newton did in order to comprehend the result and in order to avoid repeating mistakes. The concepts are somebody's concepts, Newton's, and the reasoning is his. In the narrative of the practice and the practitioner the virtues are the necessary element for the pursuit of excellence. This, of course, is true outside the confines of the practice. To understand another human action we need to know how the actor sees it and that requires sufficient knowledge of the narrative of that actor.

What is the impact on this for education for the teacher in the classroom? I have argued that when science is taught properly then, necessarily, the virtues are taught; necessarily, reflection is taught; necessarily, elements of a hard core are produced. Consequently, there would seem to be no necessity to make specific plans for the teaching of the virtues, reflection and elements of a hard core in a science lesson.

CHAPTER 19

LEARNING THE VIRTUES

> When they were about to cross to Mesopotamia, the tax-collector stationed at Zeugma took them to the registry and asked of them if they had anything to declare. 'Prudence', replied Apollonius, 'Justice, Virtue, Temperance, Courage, Perseverance,' stringing together a lot of nouns all in the feminine gender. Immediately the official, with an eye to his own profit, said, 'Well then, make me a list of your slaves.' 'I cannot,' retorted Apollonius: 'it is not my slaves I am declaring, but my mistresses.' (Philostratus, 1970)

The strands have come together and one focus is the virtues. If the argument is going to be that education involves the acquisition of the virtues and that, in teaching properly a discipline such as science, the virtues are necessarily taught, then something must be said about the virtues.

Typically, a science curriculum may say that it espouses honesty, rigour, reasoning, objectivity and justice (in the sense of doing justice to the evidence) of which at least honesty and justice are usually thought to be virtues. Perhaps, however, they are being treated as dispositions or perhaps virtues are thought to be dispositions. What is the distinction between honesty and rigour? Which are attitudes, which values and which dispositions? Also, typically, a science curriculum may suggest that the values and dispositions it enumerates are positive attributes of science and are peculiar to science. This clearly would be a mistake. The use of the word 'positive' presupposes that dispositions can be negative, in the sense of bi-polar, as can attitudes and values. The statement then asserts that honesty and the others are positive and, presumably, never negative. This creates a difficulty for the language has phrases such as 'too honest for his own good', 'what a pity he is so honest', and similar phrases in which rigour, reasoning, objectivity and justice may appear. The phrases seem to indicate that at times it is thought that these things are not a good thing but negative. They are, on occasion, a fault. However, in the English language, there are phrases which indicate that there is always an attempt to hold on to the notion that the virtues are necessarily a good thing, are positive. So 'too honest for his own good' is a way of saying a person is too candid or too outspoken. We can make sense of this by saying that the person does not have the virtue properly (Hursthouse, 1999, pp. 13-14). In this way perhaps candidness is picked out as possessing some of the qualities of honesty but it is a matter of being honest on an inappropriate occasion.

The first problem is to give some definition to the virtues. An extraordinary range characterises the generally accepted virtues. Justice, fortitude, prudence and temperance may be the cardinal virtues of antiquity but, on occasion, to them have been added compassion, tolerance, fidelity, humility, honesty, generosity, courage, simplicity and more. Comte-Sponville (2003) mentions that he started off with a list of 30 or more and whittled it down to 18 for his book. He also mentions that:

> It would be ridiculous for a treatise on virtues to strive for originality or novelty. … For the last 2,500 years, if not more, the greatest minds have thought about the virtues; my desire was to continue their efforts, in my own way and with the means available to me, using their ideas to formulate my own. (Comte-Sponville, 2003, p. 5)

I am of the same view.

What then is it important to say about the virtues in order to give an indication of what they are? First, they are not dispositions. A disposition is an inclination or a tendency which may or may not be exercised. So it may be said of someone that he or she is inclined to be or disposed to be honest and the implication is that the person chooses on this occasion to be honest but may choose not to be so in the future. Most of the time though he or she is honest. If that same person were said to be honest in the sense of possessing the virtue of honesty then he or she becomes the sort of person of whom it may be said that to be dishonest never occurred to him or her. Honesty has become part of whom he or she is. To possess the virtue of honesty is to be in the position of the exercise of that virtue being ingrained. This is not to suggest that temptation is not a possibility. It always is except, for the extraordinary.

Comte-Sponville (2003) is of the view that virtue has been thought to be an acquired disposition ever since Aristotle. Certainly Aristotle is translated as referring to dispositions when talking of the virtues (Bambrough, 1963, II 5). I take the view that virtues are not dispositions and I am doubtful that Aristotle, in the Nichomachean Ethics, was referring to dispositions as I understand them. Hursthouse takes the same view in denying that the virtues are tendencies or dispositions:

> But this is not the Aristotelian concept. Despite a few awkward exceptions (friendship, gratitude), a virtue is generally held to be a character trait, a state of one's character. If you have the virtues of, say, generosity, honesty, and justice, generous, honest and just is the sort of person you are. (Hursthouse, 1999, p. 11)

If virtues are not dispositions then what are they? I have previously mentioned MacIntyre's tentative definition in which he says that a virtue is an acquired human quality the possession of which is, on occasion, necessary to achieve certain goods (MacIntyre, 1981). The tenor of MacIntyre's remarks about virtues is in keeping with the view that virtues are character traits. The character of a person is that sum of qualities that distinguishes one person from the next. It is what he or she is. A teacher may prepare a report on a student which provides an extensive list of the student's achievements and abilities but, left at that, nothing has been said about who that student is and what may be expected of him or her apart from his or her ability to pass particular tests at a particular time. If now the teacher adds to the report that the student is honest and conscientious, loyal and trustworthy then this starts to say something about the character of the student, that is, the person that the student is. And this last list is not a list of abilities or achievements or tendencies or dispositions but more a list of expectations.

To talk of a person's character is to talk of something deep-seated in that person. A change of character is a marked change which signifies a difference which can lead observers to say that he or she is no longer the same person. For this reason it is to be expected that reliable predictions can be made about the actions of persons where their character is known. And, in the reverse direction, actions can be explained by saying the action was in character. This element of consistency, of expectation, is brought out by the phrase that an action was 'out of character'. This makes it evident why children do not have the virtues and have to acquire them. Children do not have character traits. They may well have personalities and part of that personality may be that they behave kindly towards other children but we do not think of them when very young as having character traits. These they acquire or not as they mature. Young adolescents for the most part appear to be still acquiring character traits. So they may oscillate wildly between recklessness and timidity, between compassion and indifference until maturity calms the swinging pendulum. Their characters are forming as they confront the world we live in and attempt to deal with it.

This leads to the point that the virtues are about action. They are about doing things and getting them right. As Aristotle said they are about practical wisdom or *phronesis*. A person who is as honest as the day is long is a person who possesses an excellence of character, for the virtues are excellences and their opposites are vices. The words 'virtuoso' and 'virtuosity' testify to that belief. So generosity may be contrasted with meanness on the one hand and prodigality on the other, compassion with cruelty and indifference. This brings out the Aristotelian point that the virtues typically lie in the mean between vices. The exercise of a virtue, bearing in mind the complexity of human activity, then becomes a matter of getting things right so that, for instance, there is neither recklessness nor cowardice nor foolhardiness but just the right amount of fortitude, the amount required in this situation at this time by a person such as the actor. Clearly this will be a difficult feat and will require some learning and it is by looking at how the virtues may be learnt that it is possible to add to their description.

This, it would seem, fits in with the notion that the virtues are about action which is appropriate to the circumstances. It has been argued that the virtues can also be possessed and exercised by the non-virtuous. The evil-doer can be prudent, or temperate or courageous (Comte-Sponville, 2003). It can also be argued that on occasion any virtue can be a fault (Hursthouse, 1999). Both Hursthouse and Comte-Sponville argue about individual virtues and take the results of their observations to apply to human activity. This is a mistake. Human activity is rarely, if ever, simple. It is a complex, multi-faceted process. As a result it is nearly always a mistake to think in terms of honesty being, on occasion, a fault, or, on occasion, exercised by an evil-doer. The occasion is always more complex than can be handled by the exercise of a single virtue. So, if we have in the language the phrase 'too honest for his own good' then it should be read as saying that a person lacks two or more of prudence or temperateness or justice or all of them and perhaps others. Too honest for his own good and other such phrases are phrases that we use to indicate that the mix of the virtues has not been correctly judged and applied.

Plato, in the *Protagoras*, suggests that the virtues are learnt in the same manner as a child learns his or her native language. Children learn their native language from their parents, their friends, acquaintances and the society in which they move. Apart from parents there is no one teacher of a child's native language. There are no formal lessons in a child's native language, it is a matter of listening, copying, imitating, practising, correcting, experimenting and all the other things that young children do as they seek to learn to communicate through the spoken word. In much the same way, so it may be argued, a young child learns to behave correctly. Plato is only partially right (Ryle, 1986). A young child behaving correctly is not necessarily being virtuous. Just as it is possible to use fluency and ability with language to injure so it is possible to behave in a manner in which a virtuous person would behave without having the slightest idea of virtue. And, of course, this can be extended beyond childhood for overt behaviour does not necessarily indicate virtuous action, sometimes it serves as a cloak for a vicious action. However, the point is that which Aristotle made:

> We have them [our senses], and then make use of them; we do not come to get them by making use of them. However, we do acquire the virtues by first making use of them in acts, as is also the case with techniques. Where doing or making is dependent on knowing how, we acquire the knowledge by actually doing. (Bambrough, 1963, II, 1, p. 303)

One of the commonest forms of acquiring the behaviour that goes with the virtues is training in manners. Children are taught to say 'please' and 'thank you', the first has a connection with humility – the original phrase was 'if you please' and still is in French – the second has a connection with gratitude.

> It is by mimicking the ways of virtue, that is, through politeness, that we stand a chance of becoming virtuous … Which is why it is insufficient in an adult and necessary in a child. Politeness is only a beginning but at least it is that. (Comte-Sponville, 2003, p. 11)

Another teaching method employs fairy stories and still another pithy sayings known as proverbs, 'discretion is the better part of valour' and so on. It is worth mentioning here that civilisations appear to be characterised by a collection of sayings or proverbs which set out the wisdom of that civilisation in a form that can be passed on. The young start to learn what it is to be virtuous from their elders and their acquaintances but, at the early stages, they are imitating the actions of the virtuous without being virtuous. This will come later as they come to understand and acquire the virtue that changes the perspective on their actions. But politeness can also be used as a cloak to cover viciousness and this art can be learnt in learning politeness. In the same way imitating virtuous activity can lead to vice. Practising what should be done requires that the young be placed in situations where it may not be done. So in order to practise honesty the young have to be put in a situation where they may behave honestly or dishonestly and then they are encouraged to practise the honest way of behaving. In this way they learn to be honest and acquire the virtue. But, equally, they have been put in a situation where dishonesty is a possibility and they may acquire the vice. Aristotle makes the point:

> By acting in dangerous situations, and being trained to show fear or confidence, we become brave or cowardly. The same can be said of desire and temper; some become moderate and controlled, others immoderate and bad-tempered, from actually behaving in such or such a way in cases where desire or temper is involved. (Bambrough, 1963, II, 1, p. 304)

Moral lectures, classes in moral education, the appearance of moral lessons in the curriculum can all miss the point that acquiring the virtues is a matter of learning to act in a certain way and this the young and the adult must learn by doing. Certainly the examples of others are crucial but then it is a matter of personal practice, of personal training (Ryle, 1986). It is instructive to read the above paragraphs as if they were written about science teaching. Words need to be changed here and there but the point becomes clear that being initiated into the practice of science is to be initiated into the practice of the virtues. In the light of this, the designing of specific science lessons with the aim of teaching the virtues is inappropriate. On the assumption that science teachers understand the nature of science, there are four things that are necessary:

- That science is taught properly;
- That the teachers understand that the virtues are involved in the practice of science and therefore in the learning of science;
- That the teachers understand how the virtues are involved in the practice and learning of science; and
- That the teachers understand that the virtues that are involved in science are the same virtues that are involved in social life and in the community for something is not a virtue unless it is necessary in all three.

The lesson from the *Protagoras* is applicable, as are the lessons from the ways in which we civilise the young.

It is, of course, appropriate that teachers know sufficient of the moral development of the young. The moral development of the young and adults has been a concern since time began. Recently, educational psychologists have turned their attention to the problem. Of the work that has been done, some of the most interesting comes from Kohlberg (1981, 1984). Virtue ethics may be characterised as agent-centred and concerned with how and in what way a person should come to flourish. An opposing point of view is that morality seeks to address the question of which acts are moral. Morality is about right, duty and the search for rules and principles upon which a moral agent should act. The basic premise is that an action is moral iff it follows from a moral rule or principle. That basic premise is common to both act-utilitarianism and straightforward deontological accounts. Kohlberg (1981, 1984) developed a rule-orientated account of the moral development of human beings which has connections with Kant and Piaget. Kohlberg claims that moral development is a six-stage process commencing with actions undertaken to avoid punishment or receive an award and culminating in a final stage in which there is the acceptance of and action upon a universalisable moral principle. That is, the final stage is one of Kantian transcendalism. However, if Kohlberg's thesis is considered from the point of view of a child being born into a culture then there are three broad stages in the moral development of a human being:

- The pre-cultural stage in which the person has not yet grasped the rules and procedures of the culture. The child has not grasped the conventions of the culture.
- A stage in which the person behaves in a conventional way, the conventions being determined by the culture.
- A postconventional stage in which the person is now able to make judgements on criteria which are independent of the culture and its conventions.

To the extent that the third stage presupposes criteria which are culturally independent, this is incompatible with the argument put forward in this book. The argument of this book is that there may be cultural modification or the development of a superior culture but not culturally independent criteria. Kohlberg came under attack from Gilligan (1982) who claimed his research was male-orientated. More interestingly, Gilligan argued that women thought in terms of relationships and caring and exhibited different stages of moral development to those proposed by Kohlberg. This idea can be extended and seen as raising the issue of the moral development which leads people to see themselves as inter-dependent. In virtue ethics, Hursthouse (1999) and MacIntyre (1999) have argued the importance of the interdependence of human beings and the obligations and responsibilities that arise from this. The recognition of interdependence is a mark of a flourishing human being. This points the way to an area of psychological research which it is not appropriate to follow in this book.

A virtue is an acquired character trait and it is with this point that a return to the notion of cultural science is called for. First, there is the general issue of whether or not virtues are specific to a culture so that one culture has a different list of virtues compared with another and this is a function of the difference in the cultures. Second, if it is supposed that there is such a thing as science which is idiosyncratic to a culture and thus each distinct culture has its own variant on science then, if the virtues are necessarily taught when teaching science properly, the supposition must be that a set of virtues are peculiar to a culture. If they were not then there would be a universal set of standards within and part of the idiosyncratic science of a culture and that introduces an element of incoherence. The nature of the virtues that are found in science are, it has been argued, those which are found in society for it is not a virtue if it is confined to a practice like science. Consequently, the issue of cultural relativism arises with the virtues and affects all the practices, including science, which exist in a culture. How then does one show that a standard list of virtues are necessarily to be found in any culture? The answer to this question will not only affect the general question of relativistic virtues but all practices and the formal and informal teaching of those practices. The problem here is that list is to be sought inside a culture. Hursthouse states the worry:

> The worry about such 'validation' [of the list] is that it cannot provide rational justifications at all, but is merely circular, doomed to be a mere rationalization of one's personal or culturally inculcated values. (Hursthouse, 1999, p. 165)

For MacIntyre (1981) a virtue is something that human beings may acquire and the possession of which enable human beings to achieve goods internal to practices and his argument contains a threefold test for recognising a virtue:

> He [Miller] fails to notice that, on my view, no quality is to be accounted a virtue except in respect of its being such as to enable the achievement of three distinct kinds of good: those internal to practices, those which are the goods of an individual life and those which are the goods of community. It is not that an account of justice in terms only of practices is incomplete; it is that such an account is not yet an account of justice or indeed of any other virtue. (MacIntyre, 1994, p. 284)

Hursthouse, confronting the same problem, also proposes a threefold test:

> (1) The virtues benefit their possessor. (They enable her to flourish, to be, and live a life that is, *eudaimon*.)
>
> (2) The virtues make their possessor a good human being. (Human beings need the virtues in order to live well, to flourish as human beings, to live a characteristically good, *eduaimon*, human life.)
>
> (3) The above two features of the virtues are interrelated. (Hursthouse, 1999, p. 167)

As Hursthouse defines a good human being, in part, as a good social being it would seem that she and MacIntyre are at one with respect to two of the criteria. MacIntyre has the additional criterion related to practices. However practices for MacIntyre are defined in terms of being a complex social activity. I conclude that the general tenor of their respective approaches to identifying the virtues are much the same. How may this be put into practice?

It is at this point that Hursthouse calls on Neurath and Theseus' ship. Hursthouse (1999, p. 165), quoting Quine (1950) and following McDowell (1995), accepts that there are no independent criteria to which there may be recourse and the only thing to do is to examine each putative virtue one at a time and accept, modify or reject it on the basis of reflection. It is the approach of Neurath in imagining a ship at sea replacing the planks one at a time till, finally, one may have a ship like Theseus' made of planks none of which were there at the start of the voyage. Hursthouse's reflective scrutiny is, to my mind, no different from MacIntyre's critical examination of internal coherence. This reflective scrutiny and critical examination takes place in the context of what is to count as human flourishing. This answer to cultural relativism holds good for the virtues and for the practices within a culture. It holds good against the notion of cultural science. And it holds good against constructivism of the types that were criticised earlier. In the case of the virtues as in the case of science there is always an ontological statement being made. In the case of science it is a statement about the nature of reality and the check for coherence is always in the end a check against that reality. In the case of the virtues Hursthouse makes the same point:

> Formally speaking, the concept of a virtue is specified in terms that make no reference to culture. The specification is something like: a virtue is a character trait that human beings, given their physical and psychological nature, need to flourish (or to do and fare well) – flourish *in this world in which we inevitably find ourselves, not in the particular culture or society we happen to find ourselves in.* Indeed, a fundamental criticism that can be made of a society or culture, from the standpoint of virtue ethics, is that it disallows, or makes impossible, or fails to encourage, a particular virtue in all *or* some of the human beings within it. (Hursthouse, 1995, p. 68)

It is not surprising then that the view that the virtues are necessary for human flourishing gives rise to Aristotelian virtue ethics in the West and Confucianism in the East for, in both East and West and everywhere else, what counts as and for human flourishing is always a fundamental question. The list of the virtues is an ontological statement about human nature, the world and our place in it. That list may vary from culture to culture but the list is always subject to check against reality. That reality is human flourishing. Moreover, the place of the virtues in human life can be seen to be an important argument against constructivism in the forms which I have discussed and criticized.

By way of clarification consider the matter of honesty in the practice of science. Honesty can be argued to have benefit to the individual scientist in that it is pointless and prejudicial to an individual's work to fudge results or not do justice to the hypothesis, the experiment and the results. Behaving in this way always leaves the individual open to exposure and consequent disgrace in their professional practice. There have been a number of occasions on which dishonesty has been exposed with these results. Reports of such cases are usually framed in terms of a lack of scientific integrity. It can also be argued that honesty in the practice of science enables scientists to trust each other, to rely on the work that is done, accounts that are published, to learn from each other and to advance the practice of science. It is just these sorts of consideration that Longino (1990) and Allchin (1999) argue are constitutive values of the practice of science. I do not disagree but I do wish to found it on another argument rather than leave it as an instance of values peculiar to science. Just as the virtue of honesty enables the formation and continuation of the society of scientists so the virtue performs the same function in society at large. People are able to trust one another, to rely upon another's word or directions and in this way the mortar that holds society together is produced. This is not to say that a society in order to function well must be a completely honest society. It is, as is usual in matters of the virtues, a matter of degree. However there is a limit below which the adherence to honesty must not fall if the society is to survive. Honesty benefits the practice and the practitioners and the society in which they live. It fulfils MacIntyre's three criteria and also the criteria put forward by Hursthouse. Consequently it is possible to assert with some justification that the virtue of honesty is a virtue that must appear in any standard list of virtues.

By way of further clarification consider another virtue in the teaching of science. Prudence was a virtue for antiquity and for Aquinas, but since then it has not always been regarded as a virtue. Kant certainly did not think it was. He regarded prudence as too concerned with self-interest to be a virtue. In these times we tend to think of prudence as precaution and the avoidance of harm. To call someone prudent can have an echo of criticism in that the person is unwilling to take risks. The meaning for Aristotle encompassed not only the assessment of what to avoid but also the assessment of what to choose. Human activity is complex and, in a particular circumstance, fortitude, justice, temperance, honesty and other virtues may be involved. The cardinal virtues are always involved. Prudence is that virtue which enables us to choose with foresight. This makes prudence a virtue of knowledge and reflection. Our use of language makes this plain. We say – it may be better to do this first; it would be better to take more samples if we wish a sounder base for inference; that may not happen so we had better have a fall-back position; let us think it through before trying it; if this does not work what are the options? Whenever we say or think or act on the supposition that it would be better to do this in order for X to occur then we are exhibiting prudence though nowadays we do not often use that word. All the phrases given above as examples are precisely the phrases that a science teacher may use on any number of occasions. They are also phrases that occur throughout our lives. The teacher may too often be concentrating so exclusively on guiding the student through a scientific problem or idea that he or she fails to realize or ignores the education in the virtue of prudence that may be taking place. In such circumstances the student may not acquire the virtue and, if acquired, may come to apply prudence in the science class but not elsewhere. It is important that the teacher realizes what he or she is doing and take the opportunity to invite the student to enlarge the application of the virtue and to see its place in life outside the classroom. Life is replete with situations in which people have to say to themselves that this way of going about things is not a good idea and it would be better done in another way. In this way the teacher becomes a contributor to the character formation of the student and to his or her flourishing as a human being. Where the teacher understands what he or she is doing then there is an understanding that foresight, reflection, the assessment of risk and the use of caution are being taught and, hopefully, acquired.

CHAPTER 20

WISDOM

> By three methods we may learn wisdom: First, by reflection, which is noblest; Second, by imitation, which is easiest; and third by experience, which is the bitterest. (Confucius)

For Aristotle and Confucius the fundamental question concerned human flourishing. What is it for a human being to flourish and how may this be achieved? The attempt to answer these questions has produced, over the centuries, a collection of judgements and ideas about the conduct of human life. Cultures tended to produce collections of judgements and ideas about the art of living as a flourishing human being and, in many ancient civilisations, these were collected as either an oral body of proverbs and sayings or a written literature. The proverbs and sayings were the distilled wisdom of the culture. The writings of Confucius and Mencius are concerned with human flourishing and, in that they define human flourishing and offer advice on achieving it, these writings are thought to contain wisdom. In the West the ancient Greeks produced a wisdom literature which found its classic expression in the writings of Plato and Aristotle. The Stoicism of Marcus Aurelius followed the work of the Greeks. However the wisdom literature that probably had the most wide-ranging effect on the West was that of the Jewish Old Testament and the Christian New Testament. For many centuries in the West the consideration of the nature of human flourishing and the search for wisdom were viewed as matters of supreme importance. The study of wisdom is not so viewed today.

> While modernity, at least Western modernity, has taken a rather jaundiced view of wisdom, seeing the classic quest after its meaning as a kind of fool's errand, best left to those in the business of compiling almanacs and penning messages for fortune cookies, quite the opposite was true during the centuries immediately preceding the emergence of psychology as an independent discipline, an extended period during which the study of wisdom was widely understood as the primary task of philosophers and an intellectual obligation upon all thoughtful persons. (Chandler and Holliday, 1990, p. 125)

A number of answers have been offered as to why this might be so. Taylor argues that, before the Enlightenment, the Aristotelian notion of what counted as the good life as opposed to living was accepted and that this was reversed with the Enlightenment. With the Enlightenment the core of human flourishing became production and reproduction which was to be engaged in rationally (Taylor, 1994, p. 32). Taylor's argument can be seen as accounting for the marginalisation of the pursuit of wisdom. Marcel (1951) ascribes it, the marginalisation, to the rise of science as the only recognised form of knowledge and the development of technology. The aim of the modern world is to achieve a technical dominance over the natural world. In these circumstances wisdom has fallen into disuse and then disrepute. Holliday and Chandler (1986) argue that the study of wisdom up to the fifteenth century in Europe involved examining three questions:

- what can I know?
- what ought I to do? and
- what may I hope?

They argue that gradually, from that time onwards, the first of these questions came to be dominated by the work of professional philosophers, the second became a question divorced from the first and the third became a personal matter. In the terms for which I have some sympathy, the first is dominated by the growth of science, the second becomes a matter of personal preference and the third is given no attention. Also, in terms of the individualism of modern Western society, it might be restated as; what about me? what is in it for me? and, what can I get away with? Chandler and Holliday(1990, p.125) followed their previous work by arguing that three interrelated factors may account for the modern demise of the study of wisdom:

- the tendency from the late nineteenth to the early twentieth century to equate human knowledge with scientific knowledge;
- the behaviouristic basis of much modern psychology; and
- the technological revolution.

Taylor argues to the same conclusion via a different route. With the Enlightenment came an emphasis on individual freedom and dignity. Alongside, and as part of this, came the notion that human beings should govern themselves according to reason.

> Free action was determined by rational procedures, and these strove for exactness and explicitness in order to surmount confusion and error. We have only to think of Descartes. The same spirit that pushed for an ethic of rules over one of the substantive good tended to favour an ethic of explicit rules. What was left to phronesis seemed to be abandoned to unreason. (Taylor, 1994, p. 29)

Taylor continues by commenting that the dismissal of *phronesis*, practical wisdom, is still current though the dismissal is now under attack from a number of authors.

> The study of wisdom came to be seen as metaphysical baggage.
>
> Whatever the prior status of an earlier wisdom tradition, the future shock that has accompanied the technical progress of the last century has served, according to critical theorists such as Habermas (1970), to stun us into the false expectation that all hopes for solutions to society's problems lie along a path paved with empirical facts. Because such facts would appear best assembled by a youthful breed of psychometric athletes especially skilled at inventing such technical solutions, little hope is held out for the prospect that

> elderly persons, best known for their preoccupation with a discarded past, are likely to contribute to this endeavour. Wisdom, if it has retained any meaning at all, is generally thought best understood as a kind of vestigial organ of the mind, still relevant only within those few remaining folk cultures that have not as yet joined the modern technological revolution. (Chandler & Holliday, 1990. pp. 127-128)

Chandler and Holliday are among a number of development psychologists investigating the possibility that there is human development beyond the adolescent years rather than a downward slide to incompetent old age. On the face of it this could well lead them to argue against any Piagetian or Vygotskian plateau. They argue that the instruments that have been and are used for investigating intellectual development and the interpretation of the data from those instruments are flawed and that the belief that wisdom grows with age is worth investigating. After all, it is a belief in many cultures that wisdom resides in the old and only exceptionally in the young. The phrase 'wisdom beyond her years' is token to this.

The investigation has typically involved two methods initially. The first is an investigation of what has been said about wisdom in the past before the study of wisdom fell into disuse. The second is the observation of language. A living language tends to carry with it words, phrases and meanings from its history. Just as English usage reflects an Aristotelian view of reality (compare the meaning of 'inertia' in normal usage with that in Newtonian mechanics) so that usage contains within it ideas on wisdom. The problem, of course, is that working from present usage may be to work from a usage corrupted in the demise of the study of wisdom and the analyst might also be unable to see clearly beyond present usage. Chandler and Holliday in their review of contemporary wisdom research in the field of psychology note that:

> Almost without exception the handful of investigators who have undertaken to get some fix upon the possible meaning of wisdom have followed a common course by undertaking to resurrect whatever residual knowledge may lie sedimented within the common language conceptions of ordinary persons by carrying out some form of survey of their opinions. (Chandler & Holliday, 1990, p. 130)

As they remark, the problem facing researchers at the present time is finding a model to direct their surveys and to deal with any data gathered. For instance, Kitchener and Brenner present the thesis that:

> … the research on one model of adult cognitive development, the Reflective Judgement model, illuminates the development of many attributes associated with wisdom as well as the relationship between wisdom and intelligence. (Kitchener & Brenner, 1990, p. 212)

They then go on to develop a seven stage developmental view of knowledge. Having surveyed the literature on the meaning of wisdom they suggest that many accept that wisdom can be identified and they suggest that these are the characteristics of their seventh stage of Reflective Judgement reasoning but they assert that they do not claim that their model gives a complete account of wisdom.

Having noted that psychologists are working in the field and given an indication of the methods that are being adopted, I wish to pursue the matter in a related way. The psychologists working in the field use surveys of public opinion, a philosopher may work in much the same way. in tackling the components of wisdom proceeds from the question:

> Are there any traits uniformly exhibited by the very diverse minds that by general agreement are wise? Two traits appear to stand out – reflectiveness and judgment. (Blanshard, 1967, p. 323)

This is a method equivalent to the psychologist's surveys. I shall use the other way of looking at the language.

There are numerous proverbs in the English language. Make the justifiable assumption that proverbs are pithy indications of an accumulated repository of knowledge about wisdom which has passed down to the present from many centuries and then examine some of the proverbs.

Many of them urge prudence:

- look before you leap;
- a stitch in time saves nine;
- a bird in the hand is worth two in the bush;
- don't count your chickens before they are hatched;
- many hands make light work;
- many a true word is spoken in jest;
- cast thy bread upon the waters;
- let sleeping dogs lie;
- fire is a good friend but a poor enemy; and
- he went like a lamb to the slaughter.

Others urge fortitude in the sense of moral strength:

- never say die;
- don't spoil the ship for a halfpenny of tar;
- don't cry over spilt milk;
- time and tide wait for no man; and
- the race is not to the swift nor the battle to the strong.

Others urge temperance:

- least said soonest mended; and
- empty vessels make most sound.

Others urge justice:

- he who laughs last laughs longest;
- one man's meat is another's poison; and
- spare the rod and spoil the child.

It is evident that many of the proverbs could appear in more than one category. The proverbs above are a sample of those that have survived into the modern Western world of the English language from the many that are contained in the wisdom literature generated from the time of the Ancient Greeks and the Old Testament Jewish people and the New Testament Christians. Some ancient proverbs have fallen into disuse – 'surely in vain the net is spread in the sight of the bird'; but others have been translated into a more modern idiom and then more recently fallen into disuse – 'he that spareth his rod hateth his son'. Others have been translated and survive. Still others have been enlarged – 'the race is not always to the swift nor the battle to the strong but that's the way to bet'.

The observation that can be made about all the examples of proverbs given above is that they are instances of exhortations or advice to exercise a virtue. Principally, it appears, they are exhortations or advice to exercise the cardinal virtues. This is to be expected. Necessarily a wise person would be prudent, temperate, just and have fortitude. I concede that other proverbs may relate to other virtues such as compassion, humility, honesty or benevolence but I shall confine my remarks to the cardinal virtues in the belief that what may be said about them possibly may be applied to other virtues. It is inconceivable that a wise person would not possess and exercise these virtues to a degree far higher than ordinary mortals. The cardinal virtues are intimately associated with the capacity to reflect and the exercise of that capacity. Taking a virtue ethics stance, these virtues and reflection are intimately connected with human flourishing. Therefore, in so far as wisdom is connected with the possession and exercise of the cardinal virtues, the achievement of wisdom is the pinnacle of human flourishing.

Chandler and Holliday (1990), Kitchener and Brenner (1990) and Blanshard (1967) all talk of the importance of judgement in the exercise of wisdom. It is in the judgement of which virtues to exercise and in which measure on this particular occasion that wisdom is found. Because of this, experience becomes an important asset and experience can come with age. Learning to be prudent is, in part, to learn to have foresight. Foresight depends upon the ability to predict and the ability to predict relies heavily on experience, on previous knowledge. The same form of argument applies to justice, fortitude and temperance. Consequently, the excellent exercise of these virtues depends heavily on experience. If wisdom is the exercise of these virtues then it too depends heavily on experience.

In Japan, there is embedded within the culture a set of traditional values derived from the intermingling of Confucianism, Buddhism and Shintoism. Akihiro Chiba (2004) asserts that these values may be considered positive values in modern democratic society. Confucian values are embedded in other East Asian cultures. Among the values is respect for and obedience to the elderly. The obligation and responsibility to respect the elders runs throughout Confucianism and is to be found as a value embedded in the culture of East Asia. It is not peculiar to East Asian societies. It is typical of perhaps the majority of cultures that the elders are respected. At least in part respect is to be shown because of the lengthy life experience and knowledge of the elders and the possibility of wise judgement based on this experience and knowledge.

Blanshard sums it up:

> It should be pointed out, finally, that the judgment of the wise man may carry a weight out of all proportion to that of anything explicit in his thought or

> argument. The decision of a wise judge may be implicitly freighted with experience and reflection, even though neither may be consciously employed in the case before him. Experience, even when forgotten beyond recall, leaves its deposit, and where this is the deposit of long trial and error, of much reflection, and of wide exposure in fact or imagination to the human lot, the judgment based on it may be more significant than any or all of the reasons that the judge could adduce for it. This is why age is credited with wisdom; years supply a means to it whether or not the means is consciously used. (Blanshard, 1967, p. 324)

But wisdom is not to be equated with one or other of the cardinal virtues. Wisdom is a practical thing. Aristotle called it 'practical wisdom'. And practical wisdom is exercised in confronting not the situations you have confronted before and, with your developed foresight, can handle prudently and with fortitude, justice and temperance but in confronting situations that are, to some extent novel. It is in confronting these situations that a person, if wise, realises the limits of his or her knowledge and experience, the fallibility of human judgement and, with humility and fortitude, makes a decision which, in hindsight, will have been seen to be a wise decision. There are clear links with the conclusion to which Kitchener and Brenner came having surveyed the literature.

The need to exercise wisdom arises in perplexing situations where it is difficult to predict and difficult to apply previous knowledge and experience. This is particularly so where the perplexing situation involves predictions about human action and perceptions. There is a Confucian view on the matter embedded in East Asian cultures;

> In the Chinese view, the most important factor contributing to the difficulty of predicting the future lies in the unpredictable nature of man. Thus the study of human character, through which the only hope of gaining some degree of control over future events lies, was considered a matter of vital importance to the ruler, as the present and future stability of his state often depended on his choice of ministers. This kind of study of human character which was to become from the Eastern Han onwards one of the major preoccupations of Chinese thinkers, was already of great importance in Confucius' day. (Lau, 1979, p. 22)

Children, at times, can be very perceptive in relation to a person's character, but, the norm is that the ability to assess the character of a person comes with age through which experience has been gained and, on which there has been reflection. Many members of a culture may possess the virtue of prudence which clearly involves, in social situations, the assessment of character. The prudent are not the wise though the wise possess prudence. It appears that in any culture the wise are few in number. They have the virtues of justice, prudence, temperance and fortitude. So do other members of the culture in which they exist. But these virtues are excellences and the wise possess them to an excellent degree.

It may be that with the demise of the study of wisdom and the rise of individualism in modern Western culture, Western culture has lost the respect for its elders that it once had and which East Asian culture still has, as do other cultures. The loss of that respect brings with it a loss of the value placed on experience, reflection and the study of human character. If my argument on wisdom is correct then this loss is damaging to any culture and has damaged Western culture.

It may be thought that what follows from the argument in relation to wisdom is that the teaching and learning of wisdom is not a subject which easily fits into the formal or informal learning of the child and the adolescent. This is a mistake. Certainly it is the case that there is no substitute for the experience garnered with age. However, what can be taught are those things which will enable a person to have worthwhile experience and to make use of it. Assume that the virtues and reflection will be taught because, necessarily, they are taught in the practices into which a child and a student are initiated both formally and informally as part of their education. These are the basic essentials for wisdom. What other studies should be taught both formally and informally? If the above argument is correct then it follows that at some stage children and students must study those subjects which cast light upon the character of human beings. History, for Collingwood, was rethinking the thoughts of dead men. This is a somewhat confined view but it makes the point that through a study of history it is possible to gain great insight into humanity in all its guises. Shakespeare's Hamlet is a character study of a human being caught in a situation which he might interpret in many different ways. Hamlet's problem is to find a way of seeing and understanding what has and is happening. Greene's characters in all his novels are studies in human difficulty. Owen's war poetry casts light on a particular human predicament. Literature and drama are pre-eminent as fields of study in which the human condition is exposed and explored. Macbeth, in his final hour, states the sorry view of life to which he has come. To study that final speech and trace the narrative that leads Macbeth to that position is to reflect on and gain insight into the human condition and realise the importance of three other virtues – faith, hope and charity.

None of this is to gainsay the importance of mathematics and science but it is perhaps to redress the balance. To send our youth from their schooling with the basis for wisdom is, necessarily, to equip them to flourish. The problem is that this may be incompatible with society as it is.

APPENDIX

Plan for Curricula of Nine-Year Compulsory Full-Time Primary and Secondary Schools, which became effective in September 1993. (extract)

Table 1 Values of peace and harmony embodied in syllabi of morality of primary and secondary schools

Teaching Content & Level	*Requirements of the Learners*
Filial duty to parents and respect for elders. Primary: 5 periods for Grades 1-2; 3-4 periods for Grades 3-5.	Children taught to be concerned for and considerate towards their parents, understand that listening attentively in class is an expression of respect for their teachers, and that they should willingly care for and help the elders nearby.
Solidarity and friendliness. Primary: 6 periods for Grades 1-2; 3-5 periods for Grades 3-5.	Children taught to understand that they should foster amity, sympathy, modesty, mutual help and amicable relations between schoolmates, brothers and sisters, and with neighbours.
Decorum and courtesy. Primary: 7 periods for Grades 1-2; 3-4 periods for Grades 3-5.	Children taught to understand that in China, as a land of ceremony and propriety, courtesy is a traditional virtue, to acquire its basic norms and the way one gets along with others, to respect the customs and habits of other nations, and be sedate and courteous to foreign guests.
Tolerance and kind-heartedness. Upper Primary: 4 periods for Grade 6.	Children taught: (i) to understand that it is a traditional virtue to be strict with oneself and lenient towards others; (ii) the importance of tolerance and understanding between people; (iii) to be considerate and able to think for and understand others; (iv) to show sympathy for the misfortunes of others; and (v) to forgive others for their errors.
Devotion to peace. Upper Primary: 5 periods for Grade 6.	Children taught to understand the importance of peace between people and nationalities for maintaining stability of the State and the world; people to treat others with honesty and serenity and promote peace by all efforts.
Honesty and goodwill. Junior Secondary: 4 periods for Grade 1.	Children taught to understand the importance of honesty and goodwill among people in order to ensure a good life, and to learn to treat schoolmates with honesty and sincerity.
Upholding national unity. Junior Secondary: 3 periods for Grade 2.	Children taught to be aware of provisions in the constitution concerning the duties and obligations of every citizen to help maintain the State's unity and solidarity between nationalities, and strive to uphold relationships of equality, solidarity and mutual assistance in words and actions.
Serving the State and the society. Junior Secondary: 4 periods for Grade 3.	Children taught to understand that it is the duty and obligation of every citizen to help build China into a rich, powerful, democratic, modern and civilized country, and that every citizen should exercise his or her social functions, handle the relations of interests between the individual and the State correctly, and foster the lofty ideal of selfless dedication to society.

Since 1991, the city of Yiwu has persisted in values education on peace and harmony for students of primary and secondary schools, integrating with the city's traditional virtue of love and respect for one's parents and elders. An *Outline of Guidance on Moral Education of Good Tradition in Primary and Secondary Schools of Yiwu City* was promulgated and the *Plan of Experiment on Moral Education of Good Tradition in Primary and Secondary Schools of Yiwu City* issued accordingly. Successful experiences acquired during recent years have been summed up by experimental schools and classes, some already extended for implementation in all the schools of the city. Of these, education for peace and harmony centred on filial duty and courtesy conducted by Yiwu's Experimental Primary School is a good example.

BACKGROUND TO THE LESSONS

• Filial duty and courtesy are basic moral norms in traditional Chinese culture. In ancient times, filial duty (*xiao*) was seen as the basis of loyalty (*zhong*) and a loyal minister at the time was supposedly a dutiful son. In modern times the concept of filial love and duty has been extended, oriented towards not only one's parents and loved ones, but also the motherland and the people. Education on filial duty has thus acquired a much wider social bearing. Courtesy refers to civilized acts and words as well as largeness of mind and open-heartedness. Education on courtesy aims at fostering the ideal state of mind, with which one is honest, open-hearted, strict with oneself, lenient with others, and seeking no fame or fortune.

• Filial duty and courtesy are the virtues more accessible and amenable to pupils in their school and family surroundings.

• Filial duty and courtesy are the traditional virtues cherished by people of Yiwu. The name of the city itself originates from the famous dutiful son named Yan-wu.

• Since sons and daughters of one-child families comprise the overwhelming majority of pupils in Yiwu primary schools, selfishness and over-conceitedness are quite common. The fostering of modesty and mutual consideration, therefore, constitutes a major component in classroom moral education.

THE MAIN PURPOSES OF THE LESSONS

• To foster correct notions of respecting and caring for one's parents and the habit of helping parents with household chores, doing what one can in daily life.

• To foster the habit of using polite language at any time and under any circumstances, and to strive to be amicable, modest and courteous with one's schoolmates.

WAYS AND MEANS OF CONDUCTING THE LESSONS

• *The Daily Routine of Filial Duty and Courtesy in the Experimental Primary School of Yiwu City* has been formulated, so that pupils are required to master the simple language of courtesy to be used in addressing their parents, family members, teachers and other elders, as well as their schoolmates, when they leave home, reach school, arrive at class, leave school and return home. For instance, they should say 'Good morning!' to their teachers and schoolmates when they get to school in the morning and exchange 'Good-bye!' with schoolmates when they leave school in the afternoon.

• Educational activities are organized in varied forms. For instance, the teachers tell short stories about filial love and courtesy at morning meetings or class meetings or other youth league/young pioneer-sponsored events. Or during traditional holidays such as Mothers' Day and Elders' Day, pupils are encouraged to present their parents a 'surprise' or do some good deeds for an elderly person nearby.

• Attention is drawn to 'young pacesetters' who are outstanding in filial duty and courtesy among the pupils, so that other pupils can follow the good example of their own classmates in consolidating their good behaviour.

• Schools work in close relations with the pupils' families, with teachers and parents setting themselves as examples in their conduct.

THE MAIN RESULTS ACHIEVED

• After one year's implementation of the Daily Routine, the pupils' consciousness and behaviour in respecting and caring for their parents has shown significant improvement. They have begun to understand the pains of their parents and to show care and consideration towards them by doing some trivial service; for instance, fetching a cup of tea when their father or mother arrives home, or asking them to taste some delicious food first.

• The pupils gradually cultivated the habit of using polite language appropriate for each different occasion.

• The pupils' consciousness of modesty and thoughtfulness was visibly strengthened. To test the pupils when they were confronted with an opportunity for personal gain, the school conducted a minor experiment in Class 5, Grade 3. During lunchtime one day, the class teacher deliberately prepared and brought a pot of delicious food that was a favourite of the children. However the quantity was sufficient only for ten of the children, rather than all sixteen. He told the pupils that this was so, and asked them each to take his or her own share. The sixteen children reacted calmly in taking their meal, somewhat reduced in quantity, and left some remnants in the pot. This testified to the impact, although preliminary, of education on courtesy: they had grown to think of the interests of others first.

REFERENCES

Akihiro Chiba. (2004). The evolution and development of traditional values of peace, harmony and non-violence in contemporary Japan. In UNESCO (Ed.), *Teaching Asia-Pacific core values of peace and harmony: A sourcebook for teachers.* Bangkok: UNESCO.

Allchin, D. (1999). Values in science: An educational perspective. *Science & Education, Volume 8, No. 1.*

Allen Consulting Group. (1999). *Training to compete. The training needs of Australian industry. A report to the Australian Industry Group.* North Sydney: Australian Industry Group.

Anscombe, G. E. M. (1995). Practical inference. In R. Hursthouse, G. Lawrence, & W. Quinn (Eds.), *Virtues and reasons.* Oxford: Oxford University Press.

Aquinas, S. T. (1948). *Summa theologiae (trans. Fathers of the English Dominican Province)* (I-II, Q. 61, A.2, co.). New York: Benziger Bros.

Aristotle. (1948). *Politics,* trans. E. Barker. Oxford: Oxford University Press.

Aristotle. (1984). Physics. In J. Barnes (Ed.), *The complete works of Aristotle. Revised Oxford translation.* Princeton NJ: Princeton University Press.

Ashmore, M. (1989). *The reflexive thesis: Wrighting sociology of scientific knowledge.* Chicago: University of Chicago Press.

Australian Chamber of Commerce and Industry and Business Council of Australia. (2002). *Employability skills for the future.* Canberra: Department of Education, Science and Training.

Australian Education Council. (1991). *Young people's participation in post-compulsory education and training.* Canberra: Australian Government Publishing Service.

Australian Education Council. (1992). *Putting general education to work. The key competencies report.* Canberra: Australian Government Publishing Service.

Ayer, A. J. (1964). *Man as a subject for science.* London: Athlone Press

Baines, L. A., & Stanley, G. (2000). We want to see the teacher: Constructivism and the rage against expertise. *Phi Delta Kappan, 82*(4), 327-330.

Bambrough, R. (1963). *The Philosophy of Aristotle.* New York: Mentor.

Barba, B. H. (1993). A study of culturally syntonic varibales in the bilingual/bicultural science classroom. *Journal of Research in Science Teaching, 30*(9), 1053-1071.

Barnes, B., & Bloor, D. (1982). Relativism, rationalism and the sociology of knowledge. In M. Hollis, & S. Lukes (Eds.), *Rationality and relativism.* Cambridge MA: MIT Press.

Bell, D. (2001). Communitarianism. http://plato.stanford.edu/archives/win2001/entries/communitarianism/

Benn, S. I., & Peters, R. S. (1959). *Social principles and the democratic state.* London: Allen and Unwin.

Birch, C. (1993). *Regaining compassion for humanity and nature.* Kensington, NSW: University of New South Wales Press.

Bjorklund, D. F. (2000). *Children's thinking,* 3rd edition. Australia: Wadsworth.

Blanshard, B. (1967). Wisdom. *Encyclopedia of Philosophy, 7-8.*

Bloor, D. (1981). The strengths of the strong programme. *Philosophy of the Social Sciences, 11,* 199-213.

Bloor, D. (1991). *Knowledge and social imagery,* 2nd edition. Chicago: University of Chicago Press

Boudourides, M. A. (1998). *Constructivism and education: A shopper's guide.* Paper presented at the International Conference on the Teaching of Mathematics, Samos, Greece.

Boyd, R. N. (1988). How to be a moral realist. In G. Sayre-McCord (Ed.), *Essays on moral realism.* Ithaca: Cornell University Press.

Boyd, R. N. (1991). Realism, anti-foundationalism and the enthusiasm for natural kinds. *Philosophical Studies, 61*, 127-148.

Brockman, J. (2005). Edge. www.edge.org

Bronowski, J. (1973). *The ascent of man.* London: British Broadcasting Corporation.

Brown, J. R. (1994). *Smoke and mirrors.* London: Routledge.

Campbell, N. R. (1957). *Foundations of science; The philosophy of theory and experiment.* (Formerly titled: *Physics, the elements*). New York: Dover Publications.

Celek, T., & Zander, D. (1996). *Inside the soul of a new generation.* Michigan: Zondervan.

Chalmers, A. F. (1976). *What is this thing called science?* Queensland: University of Queensland Press.

Chandler, M. J., & Holliday, S. (1990). Wisdom in a postapocalyptic age. In R. J. Sternberg (Ed.), *Wisdom, its nature, origins and development.* Cambridge: Cambridge University Press.

Chang Hao. (1996). The intellectual heritage of the Confucian ideal of Ching-Shih. In Tu Wei-Ming (Ed.), *Confucian traditions in East Asian modernity.* Cambridge, MA: Harvard University Press.

Chua, B.-H. (1995). *Communitarian ideology and democracy in Singapore.* London: Routledge.

Comte-Sponville, A. (2003). *A short treatise on the great virtues.* London: Vintage.

Conant, J. B. (1945). *General education in a free society: Report of the Harvard Committee.* Cambridge MA: The University Press.

Confrey, J. (1990). What constructivism implies for teaching. In R. Davis, C. Maher, & N. Noddings (Eds.), *Constructivist views on the teaching and learning of mathematics.* Reston VA: National Council of Teachers of Mathematics.

Couvalis, G. (1997). *The philosophy of science.* London: Sage Publications.

Cribb, J. (2006). Muzzling of science. *The Australian*, p. 26.

Daly, H. E., Cobb, J. B., & Cobb, C. W. (1989). *For the common good: Redirecting the economy toward community, the environment and a sustainable future.* Boston: Beacon Press.

Deldridge, A. e. a. (Eds.). (1987). *The Macquarie dictionary 2nd Revision.* Macquarie University NSW: The Macquarie Library Pty Ltd.

Devitt, M. (1991). *Realism and truth*, 2nd edition. Oxford: Blackwell.

Dykstra, C. (1997). Reconceiving practice in theological inquiry and education. In N. Murphy, B. J. Kallenburg, & M. T. Nation (Eds.), *Virtues & practices in the Christian tradition.* Valley Forge, PA: Trinity Press International.

Ebison, M. G. (1993). Newtonian in mind but Aristotelian at heart. *Science & Education, 2*, 345-362.

Einstein, A., & Infeld, L. (1967). *The evolution of physics.* New York: Simon & Schuster.

Foot, P. (1978a). Goodness and choice. In P. Foot (Ed.), *Virtues and vices and other essays in moral philosophy.* Oxford: Blackwell.

Foot, P. (1978b). Euthanasia. In P. Foot (Ed.), *Virtues and vices and other essays in moral philosophy.* Oxford: Blackwell.

Foot, P. (1978c). Virtues and vices. In P. Foot (Ed.), *Virtues and vices and other essays in moral philosophy.* Oxford: Blackwell.

Foot, P. (1994). Rationality and virtue. In H. Pauer-Studer (Ed.), *Norms, values, and society.* Amsterdam: Kluwer.

Foot, P. (1995). Does moral subjectivism rest on a mistake? *Oxford Journal of Legal Studies, 15*, 1-14.

Frankel, H. (1987). The continental drift debate. In H. T. Engelhardt, & A. L. Caplan (Eds.), *Scientific controversies.* Cambridge: Cambridge University Press.

Gettier, E. I. (1963). Is justified true belief knowledge? *Analysis, 23*, 121-123.

Gibbons, J. A. (1979). Curriculum integration. *Curriculum Inquiry, 9*(4), 322-332.

Gilligan, C. (1982). *In a different voice: Psychological and women's development.* Cambridge, MA: Harvard University Press.

Glasersfeld, E. v. (1987). *The construction of knowledge: Contributions to conceptual semantics.* Seaside, CA: Intersystems Publications.

Glasersfeld, E. v. (1989). Constructivism. In T. Husen (Ed.), *International encyclopedia of education: Supplementary volume.* Oxford: Pergamon.

REFERENCES

Glasersfeld, E. v. (1995a). A constructivist approach to teaching. In L. Steffe, & J. Gale (Eds.), *Constructivism in education.* New Jersey: Lawrence Erlbaum Associates.

Glasersfeld, E. v. (1995b). *Radical constructivism: A way of knowing and learning.* Washington DC: The Falmer Press.

Grene, M. (1966). *The knower and the known.* London: Faber & Faber.

Grene, M. (1985). Perception, interpretation and the sciences. In D. Depew, & B. Weber (Eds.), *Evolution at a crossroads.* Cambridge, MA: MIT Press.

Guo Xiaoping, & Song Enrong. (2004). A case study of Chinese core values on peace and harmony. In UNESCO (Ed.), *Teaching Asia-Pacific core values of peace and harmony: A sourcebook for teachers.* Bangkok: UNESCO.

Habermas, J. (1972). *Knowledge and human interests;* translated by J. J. Shapiro. London: Heinemann Educational.

Hacking, I. (1999). *The social construction of what?* Cambridge, MA: Harvard University Press.

Hardy, M. D., & Taylor, P. C. (1997). von Glasersfeld's radical constructivism: A critical review. *Science & Education, 6*, 135-150.

Haufather, S. (2001). Where's the content? The role of content in constructivist teacher education. *Educational Horizons*, (Fall), 15-19.

Hempel, C. G. (1966). *Philosophy of natural science.* New Jersey: Prentice-Hall.

Hempel, C. G. (1996). Comments on Goodman's "Ways of worldmaking". In P. J. McCormick (Ed.), *Starmaking: Realism, anti-realism, irrealism.* New York: MIT Press.

Hesse, M. (1967). Models, and analogy in science. In P. Edwards (Ed.), *Encyclopedic of philosophy.* New York: MacMillan.

Holliday, S., & Chandler, M. J. (1986). *Wisdom: Explorations in adult competence.* Basel, Switzerland: Karger.

Horton, R. (1971). African traditional thought and western science. In M. F. D. Young (Ed.), *Knowledge and control: new directions for the sociology of education.* London: Collier Macmillan.

Hursthouse, R. (1995). Applying virtue ethics. In R. Hursthouse, G. Lawrence, & W. Quinn (Eds.), *Virtues and reason: Philippa Foot and moral theory: Essays in honour of Philippa Foot.* Oxford: Oxford University Press.

Hursthouse, R. (1999). *On virtue ethics.* Oxford: Oxford University Press.

Janes, D. P. (2003). Guest editorial. *Canadian Journal of Learning and Technology, 29*(3).

Kaplan, A. (1964). *The conduct of inquiry: Methodology for behavioral sciencce.* San Francisco: Chandler for Science Research Associates.

Kaplan, A. (1999). Scientific methods in educational research. In J. P. Keeves, & G. Lakomski (Eds.), *Issues in educational research.* New York: Pergamon.

Kee, R. (1982). *Ireland: A history.* London: Sphere Books.

Keeves, J. P. (1997). Models and model building. In J. P. Keeves (Ed.), *Educational research: Methodology and measurement. An international handbook.* Oxford: Pergamon.

Keeves, J. P. (2002). Learning in schools: A modelling approach. Unpublished paper.

Kerferd, G. B. (1966). Aristotle. In P. Edwards (Ed.), *Encyclopedia of philosophy.* London: Macmillan.

Kitchener, K. R., & Brenner, H. G. (1990). Wisdom and reflective judgement: Knowing the face of uncertainty. In R. J. Sternberg (Ed.), *Wisdom: Its nature, origins and development.* New York: Cambridge University Press.

Knorr-Cetina, K. (1983). The ethnographic study of scientific work. In K. Knorr-Cetina, & M. Mulkay (Eds.), *Science Observed: Perspectives on the social study of science.* London: Sage.

Knorr-Cetina, K., & Amann, K. (1990). The fixation of (visual) evidence. In M. Lynch, & S. Woolgar (Eds.), *Representation in scientific practice.* Cambridge, MA: MIT Press.

Kohlberg, L. (1981). *Essays in moral development, Vol. 1, The philosophy of moral development.* New York: Harper and Row.

Kohlberg, L. (1984). *Essays in moral development, Vol. 2, The psychology of moral development.* New York: Harper and Row.

Kornblith, H. (1993). *Inductive inference and its natural ground: An essay in naturalistic epistemology.* Cambridge, MA: MIT Press.

Kuhn, T. S. (1967). *The structure of scientific revolutions.* Chicago: University of Chicago Press.

Kukla, A. (2000). *Social constructivism and the philosophy of science.* London: Routledge.

Lakatos, I. (1978a). *The methodology of scientific research programmes.* Cambridge: Cambridge University Press.

Lakatos, I. (1978b). Falsification and the methodology of scientific research programs. In J. Worrall, & G. Currie (Eds.), *The methodology of scientific research programs: Philosophical papers Vol. 1.* Cambridge: Cambridge University Press.

Latour, B. (1987). *Science in action: How to follow scientists and engineers through society.* Cambridge, MA: Harvard University Press.

Latour, B. (1992). One more turn after the social turn. In E. McMullin (Ed.), *The social dimensions of science.* Indiana: Notre Dame University Press.

Latour, B., & Woolgar, S. (1979). *Laboratory life: The social construction of scientific facts.* London: Sage.

Lau, D. C. (1979). *Confucius: The analects.* London: Penguin.

Laudan, L. (1981). The pseudo-science of science? *Philosophy of the Social Sciences, 11*, 173-198.

Lindsey, J. K. (1995). *Introductory statistics: A modelling approach.* Oxford: Oxford University Press.

Little, R., & Reed, W. (1989). *The Confucian renaissance.* Annandale, NSW: Federation Press.

Liu Shu-hsien. (1996). Confucian ideals and the real world. In Tu Wei-Ming (Ed.), *Confucian traditions in East Asian modernity.* Cambridge, MA: Harvard University Press.

Lombardi, O. (1999). Aristotelian physics in the context of teaching science: A historical-philosophical approach. *Science & Education, 8*, 217-239.

Longino, H. E. (1990). *Science as social knowledge: Values and objectivity in scientific inquiry.* Princeton, NJ: Princeton University Press.

Longino, H. E. (1994a). The fate of knowledge in social theories of science. In F. F. Schmitt (Ed.), *Socializing epistemology: The social dimensions of knowledge.* Lanham, MD: Rowman & Littlefield.

Longino, H. E. (1994b). Gender and racial biases in scientific research. In K. Shrader-Frechette (Ed.), *Ethics of scientific research.* Lanham, MD: Rowman & Littlefield.

MacIntyre, A. (1977). Epistemological crises, dramatic narrative and the philosophy of science. *Monist, 60*, 453-472.

MacIntyre, A. (1981). *After virtue.* London: Duckworth.

MacIntyre, A. (1988). *Whose justice? Which rationality?* Indiana: University of Notre Dame Press.

MacIntyre, A. (1990a). *Three rival versions of moral enquiry.* Indiana: University of Notre Dame Press.

MacIntyre, A. (1990b). Individual and social morality in Japan and the United States: Rival conceptions of the self. *Philosophy East and West, 40*, 489-497.

MacIntyre, A. (1991). Incommensurability, truth, and the conversation between Confucians and Aristotelians about the virtues. *Proceedings of the East-West Philosopher's Conference: Culture and modernity: East-West philosophic perspectives.*

MacIntyre, A. (1994). A partial response to my critics. In J. Horton, & S. Mendus (Eds.), *After MacIntyre.* Indiana: University of Notre Dame Press.

MacIntyre, A. (1999). *Dependent rational animals.* Illinois: Open Court.

Marcel, G. (1951). *The decline of wisdom.* London: Harvill Press.

Markman, E. (1994). Natural kinds. In H. E. Kornblith (Ed.), *Naturalizing epistemology.* Cambridge MA: MIT Press.

Markman, E., & Gelman, S. (1986). Categories and induction in young children. *Cognition, 23*, 183-208.

Markman, E., & Gelman, S. (1987). Young children's inductions from natural kinds: The role of categories and appearances. *Children Development, 58*, 1532-1541.

Matthews, M. R. (1994). *Science teaching.* London: Routledge.

Matthews, M. R. (1995). *Challenging NZ science education.* Palmerston North: Dunmore Press.

Matthews, M. R. (2000b). Constructivism in science and mathematics education. In D. C. Phillips (Ed.), *99th Yearbook*. (pp. 161-192). Chicago: University of Chicago Press.
Matthews, M. R. (2000a). *Time for science education.* New York: Kluwer Academic.
McCrindle, M. (2006). From builders & boomers to Xers and Y's: A social report on generations X & Y.
McDowell, J. (1994). *Mind and world.* Cambridge, MA: Harvard University Press.
McDowell, J. (1995). Two sorts of naturalism. In R. Hursthouse, G. Lawrence, & W. Quinn (Eds.), *Virtues and reason: Philippa Foot and moral theory: Essays in honour of Philippa Foot.* Oxford: Clarendon.
Miller, D. (1994). Virtues, practices and justice. In J. Horton, & S. Mendus (Eds.), *After MacIntyre.* Indiana: University of Notre Dame Press.
Murray, F. B. (1994). Piaget's theory. In R. J. Corsini (Ed.), *Encyclopedia of psychology.* New York: Wiley & Sons.
Association, N. S. T. (2000). Position statement on multi-cultural science education. http://www.nsta.org/positionstatement&psid=21
Nola, R. (1997). Constructivism in science and science education: A philosophical critique. *Science & Education, 6*, 55-83.
O'Neill, J. (1993). *Ecology, policy, and politics: Human well-being and the natural world.* London: Routledge.
Oakeshott, M. J. (1962). *Rationalism in politics, and other essays.* London: Methuen.
Peters, R. S. (1966). *Ethics and education.* London: Allen & Unwin.
Phillips, D. C. (1997). Coming to grips with radical social constructivism. *Science & Education, 6*, 85-104.
Phillips, D. C. (2000). Introduction. In D. C. Phillips (Ed.), *Constructivism in education: Opinions and second opinions on controversial issues.* Chicago, IL: National Society for the Study of Education University of Chicago Press.
Philostratus. (1970). *Life of Apollonius trans. by C. P. Jones.* Middlesex, England: Penguin.
Quine, W. V. O. (1969). *Ontological relativity, and other essays.* New York: Columbia University Press.
Richardson, V. (Ed.). (1997). *Constructivist teacher education: Building new understandings.* Washington, DC: Falmer Press.
Rickman, H. P. (1972). Dilthey, Wilhelm. In P. Edwards (Ed.), *The encyclopedia of philosophy.* New York: MacMillan and The Free Press.
Riegel, J. (2002). Confucius. http://plato.stanford.edu/archives/fall2002/entries/confucius/
Ryle, G. (1949). *The concept of mind.* London: Hutchinson's University Library.
Ryle, G. (1986). Can virtue be taught? In D. E. Cooper (Ed.), *Education, values and mind.* London: Routledge and Kegan Paul.
Searle, J. (1969). *Speech acts: An essay in the philosophy of language.* London: Cambridge University Press.
Searle, J. (1999). The Chinese room. In R. A. Wilson, & F. Keil (Eds.), *The MIT encyclopedia of the cognitive science.* Cambridge, MA: MIT Press.
Searle, J. (2000). *Mind, language and society: Philosophy in the real world.* London: Phoenix.
(Ed.). (2000). *The sagas of the Icelanders: The preface.* England: Viking Penguin.
Smith, R. J. (1996). The Japanese (Confucian) family: The tradition from the bottom up. In Tu Wei-Ming (Ed.), *Confucian traditions in East Asian modernity.* Cambridge, MA: Harvard University Press.
Sokal, A., & Bricmont, J. (1998). *Fashionable nonsense.* USA: Picador.
South Australian Department of Education. (2003). South Australian Curriculum Standards and Accountability Framework: Science. http://www.sacsa.sa.edu.au/index_fsrc.asp?t=About
South Australian Department of Education. (2003). South Australian Curriculum Standards and countability Framework: General Introduction. http://www.sacsa.sa.edu.au/index_fsrc.asp?t=About
Stinner, A. (1994). The story of force: From Aristotle to Einstein. *Physics Education, 29*, 77-85.

Stove, D. (1998). *Anything goes: Origins of the cult of scientific irrationalism.* Paddington, NSW: Macleay Press.

Suchting, W. A. (1992). Constructivism deconstructed. *Science & Education, 3,* 223-254.

Tattersall, I. (2002). *The monkey in the mirror: Essays on the science of what makes us.* Oxford: Oxford University Press.

Taylor, C. (1982). Rationality. In M. Hollis, & S. Lukes (Eds.), *Rationality and relativism.* Cambridge, MA: MIT Press.

Taylor, C. (1992). Atomism. In S. Avineri, & A. De-Shalit (Eds.), *Communitarianism and individualism.* Oxford: Oxford University Press.

Taylor, C. (1994). Justice after virtue. In J. Horton, & S. Mendus (Eds.), *After MacIntyre.* Indiana: University of Notre Dame Press.

Tu Wei-Ming. (1996). Introduction. In Tu Wei-Ming (Ed.), *Confucian traditions in East Asian modernity.* Cambridge, MA: Harvard University Press.

Vygotskii, L. S. (1962). *Thought and language.* Cambridge: MIT Press.

Warnock, G. J. (1967). Reason. *Encyclopedia of philosophy.*

Watt, J. (1989). *Individualism and educational theory.* Dordrecht: Kluwer.

Weber, M. (1949). *The methodology of the social sciences,* trans. Shils, E.A. and Finch, H. A.(eds). New York: The Free Press.

Wegener, A. (1924). *The origin of continents and oceans,* trans. J. Skerl. London: Methuen.

Wertheim, M. (1997). *Pythagoras' trousers.* London: Fourth Estate.

Wong, D. (2001). Comparative philosophy: Chinese and Western. http://plato.stanford.edu/archives/win2001/entries/communitarianism/

INDEX